Plant Pathogens
and the Worldwide Movement of Seeds

Edited by

Denis C. McGee
Iowa State University, Ames

APS PRESS
The American Phytopathological Society
St. Paul, Minnesota

This publication is based on presentations from a symposium entitled "Plant Pathogens and the Worldwide Movement of Seeds" held in conjunction with the annual meeting of the American Phytopathological Society on August 10, 1994, in Albuquerque, New Mexico.

This book has been reproduced directly from computer-generated copy submitted in final form to APS Press by the editor of the volume. No editing or proofreading has been done by the Press.

Library of Congress Catalog Card Number: 96-75237
International Standard Book Number: 0-89054-185-X

Printed in the United States of America on acid-free paper

The American Phytopathological Society
3340 Pilot Knob Road
St. Paul, Minnesota 55121-2097, USA

CONTENTS

Plant Pathogens
and the Worldwide Movement of Seeds

PREFATORY CHAPTER

RELEVANCE OF SEED PATHOLOGY RESEARCH PRIORITIES TO WORLDWIDE MOVEMENT OF SEED

D. C. McGee

Professor of Plant Pathology
Seed Science Center, Iowa State University
Ames, IA 50011.

Estimates of the current value of world seed market range from $40-$60 billion per year (Condon, 1997). A large sector of this market involves international movement of seeds. An important responsibility of the international phytosanitary certification system is to provide protection against the spread of plant pathogens by seeds (McGee, 1997). The past twenty years has shown major increases in the amount of commercial seed lots that cross international boundaries. US exports of seed have grown at an average annual rate of 8%, from $76 million in 1973 to over $706 million in 1995 (Condon, 1997). In addition, the International Agricultural Research Centers distribute massive numbers of germplasm and breeding line samples throughout the world annually (Diekmann, 1997). Given that seeds can be an efficient means of transmitting pathogens and that documented cases exist of major disease epidemics resulting from introduction of seed-borne pathogens (Neergaard, 1977), there is no question that precautionary measures to minimize threats from such introductions are needed. Because of the rapid expansion of global seed marketing both in developed and developing countries and the profound impact of seed-borne disease on these activities, a symposium was held at the American Phytopathological Society annual meeting in Albuquerque, New Mexico in 1994 to address the seed pathology-related interests of organizations that play a role in trading, improving, and regulating seeds in world markets. This monograph presents the views expressed by representatives of these organizations at the symposium.

Research data on seed-borne diseases generated over the past 80 years has had both beneficial and adverse effects across the range of constituencies and activities associated with international movement of seed. This prefatory chapter addresses how current research priorities in seed pathology are meeting the ideal of preventing the spread of economically important pathogens without posing unnecessary barriers to worldwide movement of seeds.

A HISTORICAL PERSPECTIVE OF SEED PATHOLOGY

In the early 20th century, seed pathology emerged as a recognizable sub-discipline from seed testing laboratories in Europe. In Germany in 1917, it was found that wheat seed infected by *Fusarium nivale* germinated well in the laboratory but expressed poor emergence in the field. A laboratory test was subsequently developed to predict field performance (Neergaard, 1977). This was the first record implicating a seed-borne pathogen as a factor

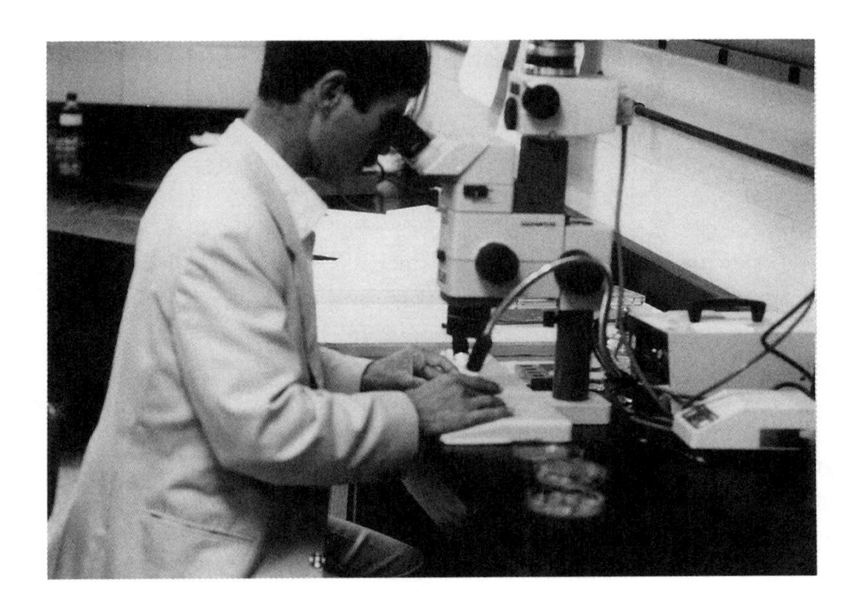

Figure 1. Participant in an ISTA seed health workshop (Courtesy of Iowa State University).

contributing to inferior seed quality. Seed health testing subsequently became an important component of seed quality analysis in seed testing laboratories throughout Europe. In 1958, the International Seed Testing Association (ISTA) formed a Plant Disease Committee (PDC) which has since held periodic workshops (Fig. 1) with the purpose of developing seed health test methods. The PDC and institutions such as the Danish Institute for Seed Pathology for Developing Countries have also provided many training courses on seed health testing. A large number of the seed health test methods used today resulted from the work of these groups. The PDC is also responsible for an extensive worldwide catalogue of seed-borne microorganisms (Richardson, 1990).

Concurrent with the development of seed health testing, epidemiological studies were conducted on important seed-borne diseases such as bacterial blights of beans, various smut diseases, and Stewart's wilt of corn

(Neergaard, 1977). Fungicide seed treatment also emerged as an important disease management tool (Jeffs, 1986). Much of this work was a component of investigations of diseases in which the seed-borne phase of the disease was considered to be of epidemiological significance. Baker (1972) was the first to identify these studies with seed pathology. He described three environments in which seeds exist: the seed production field; the period covering harvest, processing, and storing; and the planted field. He also defined categories of associations between pathogens and seeds within these environments and indicated how they related to control strategies. McGee (1981) integrated the life cycle of a plant pathogen into the three environments and suggested that the role of a seed pathologist should be to study the seed aspects of the life cycle of the pathogen and their interactions with environmental, cultural, and genetic factors that influence the cycle.

BENEFICIAL CONTRIBUTIONS OF SEED PATHOLOGY RESEARCH

Disease Management

Major reviews of seed pathology (Baker, 1972, Neergaard, 1977) cite numerous examples of successful management systems for control of seed-borne diseases. These have been achieved by cultural practices including: adjustment of planting time to facilitate disease escape; crop rotation and/or elimination of weed hosts to reduce inoculum sources; and manipulation of irrigation practices to avoid secondary infection. The elucidation of the epidemiology of Phomopsis seed decay of soybeans led to effective control of this disease by predictive methods (Fig. 2) that determine the need to apply benzimidazole fungicides (McGee, 1986, Stuckey *et al*, 1981). By

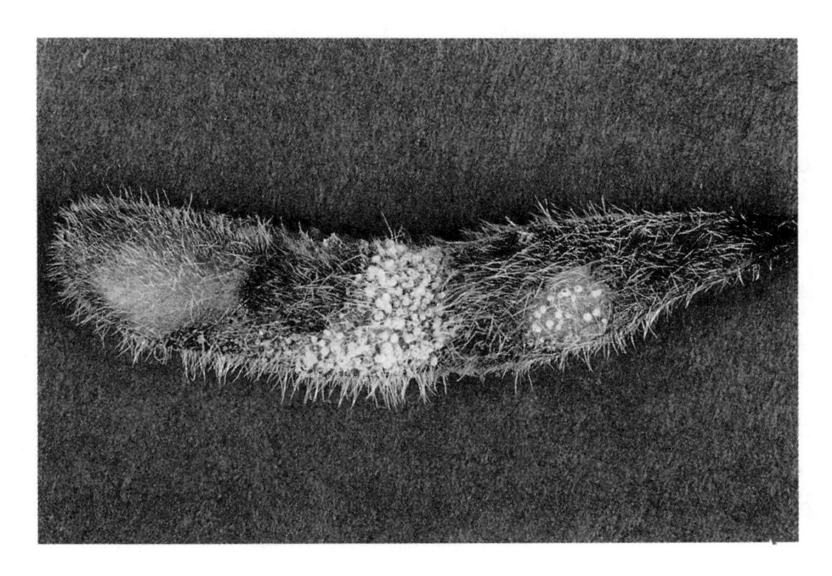

Figure 2. Pycnidia of *Phomopsis longicolla* induced on soybean pods in the laboratory test used to predict soybean seed decay (Courtesy of Iowa State University).

establishing that the potato spindle viroid was primarily pollen transmitted, control of this pathogen in potato breeding programs was accomplished by avoiding the use of pollen from infected plants (Singh *et al*, 1992). Advances in biopesticide seed treatments has occurred, in part, because attention has been paid to the mechanisms of control, particularly in the soil environment (Cook, 1993, Harman and Nelson, 1994). Inoculum thresholds of seed-borne pathogens have been applied effectively for control of economically important pathogens including *Phoma lingam* of crucifers (Gabrielson *et al*, 1977), *Pseudomonas syringae* pv. *phaseolicola* (Walker *et al*, 1964), and lettuce mosaic virus (Grogan, 1980).

Regulatory Programs

Regulatory programs based on sound epidemiological data have proven to be effective in controlling seed-borne diseases. The state of Idaho, which produces a large proportion of the snap bean seed in the United States, developed strict rules to protect the crop from bacterial blights caused by *Pseudomonas syringae* pv. *phaseolicola, P. syringae* pv. *syringae,* and *Xanthomonas campestris* pv. *phaseoli.* These rules include assurances that bean seeds imported into Idaho are free of infection by bacterial blight pathogens; fields planted in Idaho with imported seeds are inspected for foliar symptoms of bacterial blight; Idaho-produced seed is grown for two generations without bacterial blight development under surface irrigation before it can be produced under sprinkler irrigation; and fields in which bacterial blights are found are destroyed (Webster *et al,* 1983). This program utilizes knowledge of the epidemiology of the diseases and includes laboratory assays of the seeds and field disease control practices. Another effective regulatory tool is the embryo test to determine the inoculum threshold of loose smut (*Ustilago tritici*) infection of cereal. This practice is based on the finding that the incidence of loose smut infection of barley seeds was highly correlated with field disease across a range of environments and in different cultivars of barley (Rennie & Seaton, 1975). Control of seed transmission of lettuce mosaic virus can be obtained with a tolerance of 0 infected seeds in 30,000 in California, while, in the Netherlands, the tolerance is 0 in 2,000. This difference can be explained by the influence of secondary infection by aphids. The Netherlands has a cooler climate and thus a lower aphid population than does California. Furthermore, growers in the Netherlands break the disease cycle with other crops, while lettuce is grown under continuous rotation in California (Kuan, 1988). Successful elimination of pea seed-borne mosaic virus and bean common mosaic virus from US germplasm collections (Hampton *et al*, 1993,

Klein *et al*, 1990) has required sensitive and accurate techniques to assay for the pathogen in mother plants and seeds and a sound knowledge of the epidemiology of the disease throughout propagative cycles of seeds (Hampton *et al*, 1993).

ADVERSE CONSEQUENCES OF SEED PATHOLOGY RESEARCH

Misinterpretation of Catalogues of Seed-borne Diseases

The "Annotated List of Seed-borne Diseases", published by ISTA, Zurich, Switzerland, catalogues over 2400 microorganisms that have been associated with the seeds of 383 genera of plants (Richardson, 1990). This is an excellent starting point for gathering information on a seed-borne microorganism. However, because the publication does not indicate the economic importance of seed-borne pathogens, it often is misused when seed-borne microorganisms become part of phytosanitary regulations merely because they are listed in this publication. As McGee (1997) points out, this has led to a proliferation of unjustifiable phytosanitary regulations that potentially impair the movement of seeds worldwide.

Lack of Standardization in Testing Protocols

The literature is replete with descriptions of seed health test methods, but very few are thoroughly researched to determine that they are specific, accurate, reproducible, or practical. Apart from the ISTA working sheets on seed-borne diseases, there has been no systematic effort to develop standardized tests that are accepted internationally. As a consequence, seed lots sometimes are tested by different methods by the exporting and importing countries (McGee, 1997).

Impact on Seed Trading and Research Activities

Condon (1997) states that, in recent years, the US seed industry has been confronted with a host of new phytosanitary regulations, many of which cannot be scientifically justified. Countries also may require US phytosanitary certification for pests/diseases that either are known to exist within their own crop production schemes or do not exist in the United States. While lack of knowledge of the economic impact of seed-borne pathogens may underlie these types of regulations, they also can be interpreted as phytosanitary barriers to protect domestic agricultural industries. Moreover, there have been numerous incidences where US seed companies have had to return or destroy seed lots after an importing country allegedly detected a pathogen with a test method different from that used in the US phytosanitary certification process.

Dolezal (1997) points out that tight schedules for moving seeds between production areas and winter nurseries by multinational companies can be disrupted by delays associated with phytosanitary documents, or by unanticipated requirements for fungicide seed treatments.

Deikmann (1997) indicates that International Agricultural Research Centers recognize the potential for introduction of plant pathogens by seeds, and have taken strong measures in recent years minimize this problem. Nevertheless, the regulations encountered may not be totally justified and can cause enormous costs to the Center to meet requirements, as in the experience at CIMMYT in Mexico with Karnal bunt of wheat (Brennan et, 1992).

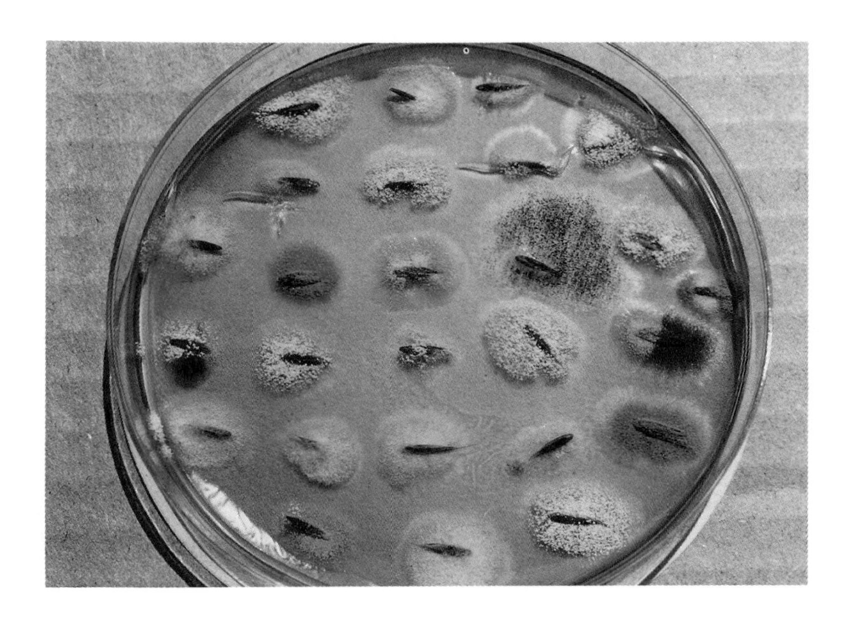

Figure 3. Several fungi growing from grass seeds in a culture plate test (Courtesy of Iowa State University).

RESEARCH PRIORITIES

A recent review of the literature on seed pathology over the period 1982-94 indicated that 23.5% of approximately 2,000 citations simply catalogued the presence of microorganisms on seeds (Fig. 3) (McGee, 1995). These types of publications are purely descriptive and do not address the potential for crop damage by planting diseased seeds or the management of seed-borne disease. Indiscriminate cataloguing of seed-borne microorganisms on seeds can obscure seed-borne pathogens that might be of genuine economic importance. A case in point is maize chlorotic mottle virus which was first reported in 1973 (McGee, 1988). No serious effort was made to determine if the pathogen was seed-borne until it caused an epidemic in winter nurseries in Hawaii (Ooka *et al*, 1990). This created concerns about importing infected corn seeds into the US. Seed transmission of the pathogen was then investigated and shown to occur (Jensen *et al*, 1991). Although cataloguing of seed-borne pathogens is necessary and is particularly important for viruses and bacteria which traditionally have been neglected due to a lack of adequate seed health assays, priority should be given to pathogens such as maize chlorotic virus mottle that meet the criteria of limited distribution and are of potential economic importance where found (McGee, 1988).

Although previous reviews of seed pathology have stressed the importance of establishing inoculum thresholds for seed-borne pathogens (Kuan, 1988, Neergaard, 1977), few citations that addressed this topic were detected in the 1982-94 literature review (McGee, 1995). Inoculum thresholds for many seed-borne pathogens are determined either arbitrarily or by field observation data (Kuan, 1988, Schaad, 1988). However, to be of value, thresholds first need to be established by experiments that include a seed health assay that is specific, accurate, reproducible, and practical. The next step is to plant seeds with different infection levels in the field and establish a correlation with subsequent plant infection. The final step in establishing an inoculum threshold is to apply appropriate statistical analysis to results. An inoculum threshold was established for *Erwinia stewartii* seed-borne infection of maize by this approach. First, an ELISA seed health assay was developed (Lamka, 1991). Laboratory and field studies then followed to relate seed infection levels to plant infection (Block, 1996). Research on inoculum thresholds is both complex and expensive. However, it is so fundamental to realistic and effective management of seed transmission of plant pathogens that little improvement in the worldwide seed health system will be possible unless priorities in seed pathology research are changed to meet this demand.

Figure 4. Pathogen extraction and dilution series steps in the seed health test for *Pseudomonas syringae* pv. *glycinea* in soybean (Courtesy of Iowa State University).

CHANGING ATTITUDES

Inconsistencies in phytosanitary regulations, their abuse as trade barriers, and increased costs to the seed industry and international research centers in obtaining phytosantitary certification has often led to mistrust between seed regulators and seed exporters. In addition, the acceptance of seed health testing methods and tolerances of seed infection that lack a credible scientific foundation has led to cynicism about the effectiveness of phytosanitary certification in protecting against the spread of the seed-borne pathogens. The requirement that soybean seeds exported from the US to the European Union (EU) be tested for *Pseudomonas syringae* pv. *glycinea* infection (Fig. 4), the cause of soybean bacterial blight, exemplifies these problems. Bacterial blight has been present on soybeans in Europe for many years (Laszlo,

1982). It is also widespread in the US but has minimal economic impact (Wilcox, 1983, McGee, 1992). US seed companies have spent approximately $1,500,000 since 1988 on testing for this pathogen on soybean seeds to meet this requirements. The seed health assay used to certify seeds has not been published or standardized. Furthermore, the pathogen must be detected in each of five subsamples of 1 kg of soybean seeds for the seed lot to be denied a phytosanitary certificate. This tolerance was determined by negotiation between US and EC authorities and has no scientific basis as a means of protecting against transmission of the pathogen by seeds (McGee, 1997).

There are, however, positive signs of change in attitude of regulators and seed industry organizations. As Condon (1997) points out, the prospects for long-term increases in world demand for seed remain very optimistic. The ultimate survivors in this extremely competitive enterprise will be those companies that develop and maintain a well-coordinated mechanism for international production and marketing, and are credible, reliable suppliers of high-quality seed. Seed health standards are being recognized as increasingly important criteria in the production of high seed quality for reasons other than to meet phytosanitary requirements. Initiatives to improve the seed health standards have come from the European and US seed industry, as evidenced by the formation of national and international committees in 1994 comprising industry and public scientists. These committees have a mandate to collaborate with regulatory authorities in the management of economically important seed-borne diseases (Wesseling, 1997, Maddox, 1997). The approach has been to assemble scientific information on seed-borne diseases, prioritize them based on economic importance, and develop management strategies that incorporate standardized assays related to currently available inoculum thresholds or those to be developed with further research.

With shrinking budgets in the public sector, governments seem willing to provide the seed industry with the opportunity for self-regulation. This has already occurred in the Netherlands, (Wesseling, 1997) and there is definite movement in that direction in the US, with the formation of the US Seed Heath Initiative (USSHI) (Maddox,1997, Klag, 1997).

CONCLUSION

Seed health issues negatively impact worldwide trading of seeds and exchange of plant breeding materials. Seed pathology is the sub-discipline within plant pathology that must address the scientific aspect of these effects. Current priorities in seed pathology research tend to focus on the traditional role of testing for pathogens in seeds. As McGee (1981) suggested, the mandate for this sub-discipline is to study the seed aspects of the life cycle of the pathogen and their interactions with environmental, cultural, and genetic factors that influence the cycle. A change in emphasis in seed pathology research to more adequately meet this mandate would lead to better protection against the spread of economically important plant pathogens without posing unnecessary barriers to international movement of seeds.

LITERATURE CITED

Baker, K. F. 1972. Seed pathology. Pages 317-416 in: *Seed Biology Vol. II*, ed. T. T. Kozlowski,. Academic Press. 447 pp.

Block, C. 1996. Biology of seed transmission of *Erwinia stewartii* in maize. Ph.D. Diss. Iowa State University, Ames, IA. 90 pp.

Brennan, J. P., Warham, E. J., Byerlee, D., and Hernandez-Estrada, J. 1992. Evaluating the economic impact of quality-reducing, seed-borne diseases: lessons from Karnal bunt of wheat. Agric. Econ. 6:345-352.

Condon, M. 1997. Implications to international trading of seeds. Pages 17-30 in: *Plant Pathogens and the Worldwide Movement of Seeds,* ed. D. C. McGee. APS Press, St. Paul, MN. 109 pp.

Cook, R. J. 1993. Making greater use of introduced microorganisms for biological control of plant pathogens. Annu. Rev. Phytopathol. 31:53-80.

Diekmann, M. 1997. Activities at the International Agricultural Research Centers to control pathogens in germplasm. Pages 41-53 in: *Plant Pathogens and the Worldwide Movement of Seeds,* ed. D. C. McGee. APS Press, St. Paul, MN. 109 pp.

Dolezal, W. E. 1997 Implications to multinational seed production. Pages 31-40 in: *Plant Pathogens and the Worldwide Movement of Seeds,* ed. D. C. McGee. APS Press, St. Paul, MN. 109 pp.

Gabrielson, R. L, Mulanax, M. W., Matsuoka, K., Williams, P. H., Whiteaker, G. P., and Maguire, J. D. 1977. Fungicide eradication of seed-borne *Phoma lingam* of Crucifers. Plant Dis. Rep. 61:118-121.

Grogan, R. G. 1980. Control of lettuce mosaic with virus free seed. Plant Dis. 64:446-449.

Hampton, R. O, Kraft, J. M, and Muehlbauer, F. J. 1993. Minimizing the threat of seed-borne pathogens in crop germplasm: elimination of pea seedborne mosaic virus from the USDA-ARS germ plasm collection of *Pisum sativum*. Plant Dis. 77:220-224.

Harman, G. E. and Nelson, E. B. 1994. Mechanisms of protection of seed and seedlings by biological seed treatments: implications for practical disease control. Pages 283-292 in: *Seed Treatments: Progress and Prospects,* Mono. 57, ed. T. Martin, BCPC, Thornton Heath, UK. 482 pp.

Jeffs, K. A. 1986. A brief history of seed treatment. Pages 1-5 in: *Seed Treatment 2nd ed,* ed. K. A. Jeffs. British Crop Protection Council, U.K. 332 pp.

Jensen, S. G, Wysong, D. S, Ball, E. M, and Higley, P. M. 1991. Seed transmission of maize chlorotic mottle virus. Plant Dis. 75:497-498.

Klag, N. G. 1997. Seed Quarantine system in the United States. Pages 55-66 in: *Plant Pathogens and the Worldwide Movement of Seeds,* ed. D. C. McGee. APS Press, St. Paul, MN. 109 pp.

Klein, R. E, Wyatt, S. D, and Kaiser, W. J. 1990. Effect of diseased plant elimination on genetic diversity and bean common mosaic virus incidence in *Phaseolus vulgaris* germplasm collections. Plant Dis. 74:911-913

Kuan, T. L. 1988. Inoculum thresholds of seed-borne pathogens: overview. Phytopathology 78:867-868.

Lamka, G. L., Hill, J. H., McGee, D. C., and Braun, E. J. 1991. Development of an immunosorbent assay for seed-borne *Erwinia stewartii*. Phytopathology 81:839-846.

Laszlo, E. M.. 1982. Infection of soyabean varieties by bacterial blight (*Pseudomonas glycinea* Coerper) under field conditions. Novenyvedelem 18:443-447.

Maddox, D, A. 1996.Regulatory Needs for Standardized Seed Health Tests. Pages 81-92 in: *Plant Pathogens and the Worldwide Movement of Seeds,* ed. D. C. McGee. APS Press, St. Paul, MN. 109 pp.

McGee, D. C. 1981. Seed pathology: its place in modern seed production. Plant Dis. 65:638-642.

McGee, D. C. 1986. Prediction of Phomopsis seed decay by measuring soybean pod infection. Plant Dis. 70:329-333.

McGee, D. C. 1988. *Maize Diseases: a Reference Source for Seed Technologists.* APS Press St. Paul, MN. 150 pp.

McGee, D. C. 1992. *Soybean Diseases: a Reference Source for Seed Technologists.* APS Press St. Paul, MN 151 pp.

McGee, D. C. 1995. An epidemiological approach to disease management through seed technology. Annu. Rev. Phytopathol. 33:445-466.

McGee, D. C. 1997. World phytosanitary system: problems and solutions. Pages 67-109 in: *Plant Pathogens and the Worldwide Movement of Seeds,* ed. D. C. McGee. APS Press, St. Paul, MN. 109 pp.

Neergaard, P. 1977. *Seed Pathology, Vols. I, II.* John Wiley & Sons, NY. 1187 pp.

Ooka, J. J., Lockhart, B. E., and Zeyen, R. J. 1990. New maize disease virus in Hawaii. (Abstr.) Phytopathology 80:892.

Rennie, W. J. and Seaton, R. D. 1975. Loose smut of barley: the embryo test as a means of assessing loose smut infection in seed stocks. Seed Sci. Technol. 3:697-709.

Richardson, M. J. 1990. *An Annotated List of Seed-borne Diseases 4th ed.* International Seed Testing Association, Zurich, Switzerland.

Schaad, N. W. 1988. Inoculum thresholds of seed-borne pathogens: bacteria. Phytopathology 78:872-875.

Singh, R. P., Boucher, A., and Somerville, T. H. 1992 Detection of potato spindle tuber viroid in the pollen and various parts of potato plant pollinated with viroid-infected pollen. Plant Dis. 76:951-953.

Stuckey, R. E., Jacques, R. M., Tekrony, D.M., and Egli, D. M. 1981. Foliar fungicides can improve seed quality. Kentucky Crop Improvement Association, Lexington, Kentucky. 8 pp.

Walker, J. C., and Panel, P. N. 1964. Splash dispersal and wind factors in epidemiology of halo blight of bean. Phytopathology 54:140-141.

Webster, D. M., Akin, J. D., and Cross, J. E. 1983. Bacterial blights of snap beans and their control. Plant Dis. 67:935-940

Wesseling, J. H. 1997. Government/Vegetable Seed Industry cooperation in the Netherlands. Pages 93-103 in: *Plant Pathogens and the Worldwide Movement of Seeds,* ed. D. C. McGee. APS Press, St. Paul, MN. 109 pp.

Wilcox, J, R. 1983. Breeding soybeans resistant to diseases. Pages 183-235 in: *Plant Breeding Reviews. Vol. 1.* ed. J. Janick AVI Publishing Co. Westport.

CHAPTER 1

IMPLICATIONS OF PLANT PATHOGENS TO INTERNATIONAL TRADING OF SEEDS

M. S. Condon

Vice President of International Marketing
American Seed Trade Association
601 13th St, N.W., Washington, D.C. 20006

In order to foster a common understanding of international trading of seeds, it is beneficial to begin from a US seed industry perspective. The magnitude of potential growth in this area is sizable and the importance of trade to the US seed industry cannot be understated. By examining the history of international seed trade we gain insight into its evolution and environment. Reviewing trends in phytosanitary regulatory action implemented by contemporary plant health officials throughout the world will lead to a better understanding of the implications of plant pathogens and their impact on the international movement of seeds.

OVERVIEW OF THE US SEED TRADE

Although not supported by hard data, estimates of the current value of world seed market range from $40-60 billion. The US market accounts for about 10% of this total. The leading five multinational seed companies sell $3-4 billion or only 4-5% of the world market value. These figures indicate that there is considerable room for expansion of the US seed industry into the international market.

Statistics compiled by the USDA Foreign Agriculture Service indicate that US exports of seed grew at an average annual rate of 8% from $76 million in 1973 to over $706 million in 1995 (USDA/FAS, 1996). Seed imports have also been growing, but at a lesser rate, and now stand at approximately $290 million. Reasons for growth in US seed exports include:

- *The US seed industry has been innovative and aggressive in pursuing new markets worldwide.*
- *The decline in domestic market share for various seed types and varieties has encouraged many US seed companies to include the international market as an integral component of their business operations.*
- *Increased global economic interdependence and competition among foreign seed sectors, and corporate profitability often hinges on global market expansion.*
- *A high degree of technological sophistication in the development of new varieties and the production of quality seed continues to be the main ingredient for past and future success in US seed trade.*
- *The climatic and geographic diversity of the United States better assures adaptability of US seed varieties to different climate and soil conditions worldwide and allows a comparative advantage in the production of a number of seed species and varieties.*

- *There is a growing appreciation and need worldwide for the high genetic and physiological quality of seed varieties produced in the United States.*

Given these advantages, together with recent trends in US seed exports, it is conceivable that US seed exports will reach the $1 billion mark in the foreseeable future (Tables 1 and 2).

Export sales and development of international operations have been significant factors supporting company growth within the US seed sector during the last two decades. More than 150 companies in the US seed industry currently trade seed internationally. Collectively, these companies conduct business with over 95 countries throughout the world. Despite the high level of world market penetration, enormous trade opportunities still remain. For example, the former Soviet Republics, Eastern Europe, India, China, Iran, Vietnam, Sub-Saharan Africa,

Table 1. US seed exports by selected years and geographic regions ($ millions)[a]

Geographic Area	1994-95	1993-94	1992-93	1991-92
North America	183.7	189.0	190.6	190.2
South America	55.8	50.0	41.3	30.5
European Union	238.3	191.0	223.0	240.0
Other Western Europe	9.6	6.4	6.4	9.8
Africa	17.5	21.0	19.7	15.8
Asia	97.0	76.0	74.8	63.9
Oceania	13.6	11.5	8.5	9.6
Caribbean/ C. America	10.3	13.0	11.2	11.5
CIS/Eastern Europe	9.5	31.6	41.1	4.2
Middle East	38.8	32.5	53.9	63.5
TOTAL	674.1	622.0	670.5	639.0

[a]Source: USDA/FAS 1994. US Planting Seed Trade" FAS/USDA, Circular Series FFVS 2/94 and USDA/FAS 1995. US Planting Seed Trade" FAS/USDA, Circular Series FFVS 4/95.

Table 2. US seed exports by selected years and product group ($ millions)[a]

Product Group	1994-95	1993-94	1992-93	1991-92
Forage-Turf	123.5	111.6	121.8	107.7
Vegetables	244.1	232.8	224.8	217.6
Flowers	24.1	20.7	19.3	15.7
Corn, Field	179.6	164.8	189.7	164.0
Grain Sorghum	11.7	25.5	31.0	40.3
Sunflower	21.4	9.8	10.1	10.0
Soybean	18.5	22.3	28.3	35.8
Sugar Beet	2.9	3.6	3.0	n/a[b]
Tree and Shrub	1.4	1.9	1.9	n/a
Wheat	1.7	6.9	49.5	n/a
Other	12.0	14.6	10.1	17.2
TOTAL	640.9	614.5	689.5	591.1

[a]Source: USDA/FAS 1994. US Planting Seed Trade" FAS/USDA, Circular Series FFVS 2/94 and USDA/FAS 1995. US Planting Seed Trade" FAS/USDA, Circular Series FFVS 4/95.
[b]Data not available

Brazil and Australia continue to buy relatively little seed from external sources. In addition, as Table 3 illustrates, the top 10-11 countries that imported US seed varieties from 1991-94 accounted for 72% of total US exports (USDA/FAS, 1995). The international seed trade is a very dynamic enterprise in which existing varieties are being replaced by improved varieties on average every 3-5 years to meet the demands of changing farming systems worldwide. There are currently over 52,000 varieties of seed produced in the United States, each with unique marketable characteristics and many of which are adaptable to foreign markets. Advances in biotechnological methods, seed treatments, seed coatings, and processing technology also continue to improve and enhance the marketability of US seed overseas.

Table 3. US seed exports by selected years for top 14 countries ($ millions)[a]

Country	1994-95	1993-94	1992-93	1991-92
Mexico	104.0	109.7	112.6	116.7
Japan	60.0	53.1	57.2	47.7
Italy	67.8	62.0	84.1	98.0
Canada	79.7	79.3	78.0	73.5
Netherlands	42.4	31.9	31.8	37.8
France	43.0	38.3	48.7	38.4
United Kingdom	n/a[b]	n/a	n/a	10.8
Saudi Arabia	16.7	14.4	52.5	64.6
Germany	n/a	n/a	11.3	13.6
Greece	n/a	15.8	n/a	9.0
Spain	22.1	14.7	12.7	17.3
Argentina	28.3	22.4	21.3	n/a
Romania	n/a	n/a	12.1	n/a
Hong Kong	15.8	n/a	n/a	n/a
TOTAL	479.8	441.6	522.3	527.4

[a]Source: USDA/FAS 1994. US Planting Seed Trade" FAS/USDA, Circular Series FFVS 2/94 and USDA/FAS 1995. US Planting Seed Trade" FAS/USDA, Circular Series FFVS 4/95.
[b]Data not available

In a very competitive trade environment, seed companies must have access to an ever improving genetic base for the breeding and production of higher-quality, higher-yielding varieties for global markets. They must also establish themselves as well-known and reliable suppliers of seed with importers and government officials of importing countries if they are to compete effectively in the long term.

Increasing seed production costs in many industrialized countries has made it necessary for a number of seed companies to establish seed production facilities in countries where land and labor costs are relatively low. Moreover, the rapid pace by which new varieties are being developed and distributed has also necessitated that companies utilize winter breeding and production facilities to accelerate the introduction and distribution of improved varieties in the world market. Consequently, seed has become a commodity subject to very sophisticated international production, processing, and distribution systems.

There is no doubt that the US seed industry will continue to aggressively pursue new trade opportunities to increase and preserve their global market share well into the future. Failure to respond to international trade opportunities will both diminish the role of the US seed industry in the international marketplace and allow foreign competitors to dominate agricultural sectors worldwide, including the United States. It is clear, therefore, that the international market for seed is a critically important and indispensable component of the US seed industry in terms of its viability and continuity.

HISTORIC OVERVIEW OF INTERNATIONAL TRADE

At this juncture, it is useful to explore the historic evolution of international seed trade to gain a better understanding of the evolution and environment in which international seed trade is conducted. McMullen (1987) in his book "Seeds and World Agricultural Progress" confirms that "privatization" or "commercialization" of international seed trade is a recent phenomenon. It has only been during the past thirty years that the world has experienced dramatic emergence of private seed companies in the global marketplace. This is contrasted with the relatively recent decline in the number of public institutions involved in seed research and production. For the most part, public institutions continue to concentrate on basic research and/or the production of seed species that are strategically important to national and international food security. However, many government seed institutions that previously carried out seed research have had their activities limited to regulatory programs for both domestic and international seed trade. These trends have even taken root, to some degree, in the middle income and developing countries.

As the world's private seed sector continues to evolve and grow, it is expected that the influence of public research and regulatory institutions will continue to decline. A number of domestic and international government seed regulatory programs may be passed on to the private sector as public funds decline and the international seed industry demonstrates willingness and ability to become self-regulating. This has already happened in the Netherlands (Wesseling, 1997).

McMullen (1987) also points out that it was not until the 1970's, with the advent of adaptable high-yielding varieties for global markets, that the world seed industry began to recognize the potential growth in international trade. The late 1970's and 1980's also saw extensive acquisitions of or mergers with seed companies by large companies interested in expanding the international sector of their business. This period of global commercialization precipitated among other things:

- *Increased dependence on trade and foreign suppliers of seed.*
- *Barriers to trade and regulations on production established by countries to keep foreign competition in check .*

With wider acceptance of plant variety protection legislation by foreign governments and its favorable impact on research and development, the number of seed suppliers entering the international marketplace are expected to increase. Seed companies that will have an advantage in this extremely competitive environment are those that develop and maintain a well-coordinated international production and marketing mechanism, an adaptable research base, and maintain credibility as a reliable supplier of high-quality seed.

The prospects for long-term increases in world demand for seed remain very optimistic as continued growth in the world's population guarantees increased demand for high-quality, high-yielding seed varieties. This increased demand will also be maintained by the continued development and distribution of improved varieties. In an effort to remain competitive in the global economy as well as to ensure food self-sufficiency, middle income and developing countries are expected to increase their use of hybrids and other improved varieties.

In the interim, however, international trade in seed will remain volatile and seed companies will come in and out of being. Margins for seed are expected to remain modest in view of the high level of competition existing in the international marketplace.

THE IMPACT OF SEED-BORNE DISEASES ON INTERNATIONAL SEED TRADE

Trade policy experts indicate that phytosanitary issues, including seed health, will be at the forefront of future international agricultural trade disputes. As countries gradually reduce import quotas and tariffs on agricultural commodities as a result of GATT and other trade liberalization agreements, they will increasingly utilize phytosanitary barriers as a fail-safe measure to protect their domestic agricultural industries. Seed has not and will not be exempt from this trend.

In recent years, the US seed industry has been confronted with a host of new phytosanitary regulations, many of which cannot be scientifically justified. Two such regulations that have had major economic impact on US seed producers are the requirements to test for *Pseudomonas syringae* pv. *glycinea* in soybean seeds

Figure 1. Stunting and leaf streaking symptoms on corn plants severely infected by *Erwinia stewartii*. (Courtesy of CIMMYT).

exported to the European Community and for *Erwinia stewartii* (Fig. 1) in maize seeds exported to over 70 countries worldwide. Credible scientific information exists to indicate that seed transmission of these pathogens present minimal economic risk to importing countries (Block, 1996, McGee, 1992). Unjustified regulations often are implemented because regulators in the importing-country lack knowledge of the economic risks of introducing particular seed-borne pathogens. Countries have also been known to require US phytosanitary certification for pests/diseases that either are known to exist within their own crop production schemes or do not exist in the United States. Sorghum seeds exported from the US to Mexico have to be tested for *Pseudomonas andropogonis* despite the fact that *P. andropogonis* is common in all of the important production areas of Mexico (Narro *et al*, 1992). Mexico also has a regulation that pea seeds be tested for *Corynebecterium flaccumfaciens* pv.

flaccumfaciens although peas have not been recorded as a host for this pathogen. Where phytosanitary requirements can be met by US seed exporters, the shipment may also require import permits. These often are issued after seed health inspection and certification are completed, creating a further obstacle to movement of the seeds. Consequently, US seed exporters are often prohibited from shipping seed to a number of countries and many US seed varieties become less competitive through increased costs associated with excessive phytosanitary controls.

The US has extensive scientific literature on plant diseases and conducts regular surveys for new and economically important pathogens. As a consequence, the US is vulnerable to excessive regulation by countries that have limited resources to determine the pathogens present in their countries. The technological strength of the US seed industry in this area can be an "Achilles heel". The extensive information base on seed-borne diseases and plant pathogens helps maintain the quality of US seed, but also allows foreign competitors to utilize this information to protect their domestic industries through the issuance of prohibitive phytosanitary import requirements for US seed. An example is a report that *Pseudomonas syringae* pv. *glycinea* caused extensive losses in soybeans in Iowa in the 1970's (Kennedy and Alcorn, 1980). Although the authors later stated that the estimate of losses was greatly exaggerated (D. C. McGee, *pers. com.*), EU authorities continued to use the report as justification for the regulation for that pathogen.

The majority of seed health tests used throughout the world have never been subject to standardization that would ensure accuracy and repeatability. Apart from a limited number of recommended tests for seed-borne diseases issued as working sheets by the International Seed Testing Association, there are few seed health tests that meet defined international standards. There have been numerous incidences where US seed companies have

Figure 2. Zonate leaf spot of sorghum, caused by *Gloeocercospora sorghi* (Courtesy G. Odvody).

had to return or destroy seed lots after an importing country allegedly detected a pathogen with a test method different from that used in the US phytosanitary certification process.

Historically, there is no question that some diseases have been introduced into countries by seeds and that phytosanitary regulations are necessary. Examples include, *Gloeotinia temulenta*, the cause of blind seed of ryegrass, that was introduced on seeds to Oregon, US from New Zealand in 1940 (Hardison, 1948) and the occurrence of zonate leaf spot (*Gloeocercospora sorghi*) (Fig. 2) in Venezuela that was attributed to imports of sorghum seeds from the US (Ciccarone, 1949). However, the ability of the

US seed industry to counter the disproportionate amount of unfair phytosanitary policies and regulations which limit seed trade will be tested in the coming years.

CONCLUSION

The great advances made by the US seed industry over the past three decades may be in serious jeopardy if the current trend in excessive phytosanitary regulation is not controlled. If seed is not allowed to move freely and unencumbered by excessive and unfair phytosanitary regulation, the merits of high-yielding and high-quality US seed may not be realized by many in the global agricultural community. The extreme competitiveness of the world market for seed and costs associated with excessive phytosanitary regulation often make US-produced seed non-competitive, unavailable, and beyond the reach of farmers throughout the world. In view of the extent of world market penetration achieved by the US seed industry, the looming phytosanitary crisis may have serious consequences for countries dependent on the availability of improved plant genetics in the years ahead.

One problem is that as governments worldwide transfer responsibilities from a public to a private seed sector, there is a tendency to overregulate the private seed sector. It also is evident in negotiations with plant health officials worldwide that foreign public plant health officials generally mistrust the intentions and motives of the private seed trade. These plant health officials are usually sincere about protecting their domestic agriculture from the introduction of economically harmful pests, but they may have a poor understanding of potential economic losses or actual risks associated with the transmission of particular pests and diseases. More importantly, there is limited understanding of the economic consequences associated with phytosanitary regulations in terms of the price and marketability of seed for particular farming systems.

Considering that "commercialization" of international seed trade is a recent and rapidly expanding phenomenon, it is no surprise that seed regulatory programs in many countries have not kept pace. In numerous instances, national seed programs have only recently turned their focus from research and production of domestic seed supplies to regulating the private seed traders conducting business within their country. With more companies entering the international market, increasing sophistication of global production and distribution systems, and increasing privatization of seed sectors, there has been an almost frantic effort by plant health official worldwide to gain control over the surge in private trade.

Unfortunately, limited priority has been given to the scientific study and control of exotic seed-borne pathogens by foreign plant health officials. Consequently, the common procedure employed in establishing a country's seed health regulations has been to extensively regulate all pathogens common to a particular plant species, seed-borne or not, and have the US plant health officials and/or US seed industry provide scientific evidence supporting the elimination of particular seed health requirements.

Lastly, the prevalent policy of many middle income and developing countries throughout the world has been to achieve self-sufficiency in seed production through the protection/support of domestic seed industry and to import only that which cannot be produced internally. This policy continues to give rise to most of the trade protection measures, including phytosanitary restrictions, encountered by the US seed industry.

In closing, we can clearly conclude that seed-borne diseases have a significant impact on the world seed trade. The solution to the problem can only be achieved through greater cooperation and coordination between US plant health officials, the US seed industry and other foreign governments and seed industries in the systematic organization and dissemination of scientific information on seed-borne diseases.

LITERATURE CITED

Block, C. C. 1996. Biology of seed transmission of *Erwinia stewartii* in maize. Ph.D Dissertation, Iowa State University. 90 pp.

Ciccarone, A. 1949. Zonate leaf spot of sorghum in Venezuela. Phytopathology 39:760-761.

Hardison, J. 1948. Blind seed disease of perennial ryegrass. Oregon Agric. Exp. Stn. Circ. 177, 11 pp.

Kennedy, B. W., and Alcorn, S. M. 1980. Estimates of U.S. crop losses to procaryote plant pathogens. Plant Dis. 64:674-676.

McGee, D. C. 1992. *Soybean Diseases: a Reference Source for Seed Technologists.* APS Press St. Paul, MN 151 pp.

McMullen, N.1987. *Seeds and World Agricultural Progress. Report No. 127.* National Planning Association.

Narro, J., Betancourt, V. A,, and Aguirre, J. I. 1992. Sorghum diseases in Mexico. Pages 75-84 in: *Sorghum and Millet Diseases: a Second Review.* eds W. A. J de Milliano, R. A. Frederiksen, and G. D. Bengston. International Crops Research Institute for the Semi-Arid Tropics. 370 pp.

USDA/FAS 1994. US Planting Seed Trade FAS/USDA, Circular Series FFVS 4/94.

USDA/FAS 1995. US Planting Seed Trade FAS/USDA, Circular Series FFVS 2/95.

USDA/FAS 1996. US Planting Seed Trade FAS/USDA, Circular Series FFVS 13/96.

Wesseling, J. B. M. 1997. Government/Vegetable Seed Industry cooperation in the Netherlands. Pages 93-103 in: *Plant Pathogens and the Worldwide Movement of Seeds,* ed. D. C. McGee. APS Press, St. Paul, MN. 109 pp.

CHAPTER 2

IMPLICATIONS TO MULTINATIONAL SEED PRODUCTION

W. E. Dolezal

Research Plant Pathology Coordinator
Pioneer Hi-Bred International Inc.
PO Box 1004, Johnston, IA 50131

Higher crop yields and increased stability of yield is a common goal of many agricultural scientists. Grain yields of maize in the United States during the thirty year period from 1955-1985 increased by an average of 133 kg/ha per year (Duvick, 1989). This increase can, in part, be attributed to control of plant disease. Disease control may be achieved by cultural practices, application of pesticides or by deployment of resistant varieties developed in plant breeding programs. Many plant breeding programs require the international movement of useful germplasm by seed, which comes with the risk of accidental introduction of a plant pathogens. This paper will deal with management strategies used in a major hybrid maize seed breeding program to avoid such introductions and relates particularly to shipments of seeds to and from winter nursery locations in subtropical environments. The views expressed are those of the author, drawn from fourteen years of experience in dealing with phytosanitary issues for a multinational commercial seed company.

STRATEGIES FOR AVOIDANCE OF SEEDBORNE PATHOGEN INTRODUCTIONS

Role of Company Plant Pathologists

Accidental introductions of pathogens and insects into countries that cause economic damage are a real hazard in the international movement of seeds. Several seed-borne downy mildew pathogens of corn that are restricted in distribution to southeast Asia are considered to be major economic threats to other maize production areas of the world (Thurston, 1973). There currently is great concern regarding the potential economic impact of the recent discoveries of Karnal bunt (*Tilletia indica*) infection of wheat in southwestern US and ergot of sorghum (*Sphacelia africana*) (Fig. 1) in Central American countries and the US. These examples serve to remind pathologists to remain vigilant in avoiding these problems.

Multinational seed companies rely, in part, on their plant pathology staff to provide guidelines for producing the best quality seed. Pathologists also advise the crop production staff of changes in phytosanitary regulations in countries where business is conducted. They recommend precautionary measures to eliminate the possibility of introductions of exotic pathogens in the shipping of seeds between international research locations. This can be a very complicated exercise. Unintended introductions of

Figure 1. Ergot on sorghum inflorescences (Courtesy of ICRISAT).

exotic pathogens into a new area could result in severe economic loss to agricultural businesses in the affected region and the company or institution responsible for the introduction could also be subject to costly civil lawsuits. Equally destructive are associated losses of customers and of business and research reputations. Pioneer Hi-Bred International, Inc. currently has 63 maize research locations throughout the world and each country has specific phytosanitary requirements which must be followed. Therefore, it is important to know which pathogens are considered to be of quarantine status. McGee (1988) published a list of the diseases of maize that have been cited in literature as being seed-borne.

Seed Treatment Regulations

Fungicide seed treatments may be required by a country before seed is allowed entry. For example, Mexico mandates this for several types of seeds imported from the US. Additional problems arise when a class of fungicides may be approved for use by the importing country, but it may be illegal for use as a seed treatment in the exporting country. Several ergosterol biosynthesis inhibitor- based fungicides are registered for use as cereal seed treatments in Europe, but not in the US. Under these circumstances, alternative fungicides with similar efficacy may be substituted after negotiation. Therefore, it is important to ascertain these requirements before seed is ready to be sent to the winter nursery.

Generation Cycle Time

Reduction of the time period between generations of seeds is vitally important to maize plant breeding programs. Yield on any given test cross hybrid is collected in the fall. Inbred parents that have been identified as having high general combining ability are harvested from the summer nursery, shelled, packaged, and shipped to winter nurseries, many of which are in other countries. At the winter nursery, the inbred will be selfed or crossed onto other test cross parents to make new hybrid combinations that will be tested for yield in the following summer. Each winter nursery location may have from six to twelve different planting dates for increasing seed. Winter nursery timetables must be followed precisely if seed is to arrive at the research station in time to meet planting schedules in the following spring. Any delay with custom officials in clearing the winter nursery seed can, therefore, be very costly to a breeding program. Winter nursery managers and pathologists must have all necessary phytosanitary documents ready to accompany each seed shipment. They must also work with local custom officials to assure that proper notification is given for expected seed arrival dates at the port of entry.

Phytosanitary Requirements for Winter Nursery Seed Shipments

Seeds usually are visually inspected upon arrival at a port of entry and, if no pathogens are detected, they are released for planting. However, the local regulatory agency may decide to draw random samples for more detailed laboratory testing for the presence of seed-borne pathogens. It is common to sample small sub-lots from large bulk shipments of commercial hybrid seed. However, it is a much more complicated matter to sample seeds in winter nursery (Fig. 2) shipments. These may comprise thousands of small seed packets, each containing only twenty kernels. These packets represent hundreds of genotypes from many different locations. There is no uniformly accepted system for handling this type of seed shipment among countries. Special phytosanitary requirements may be established between regulatory agencies in exporting and importing countries, but officials at various ports of entry to a country may not be aware of them. This not only creates delays in the entry of winter nursery shipments, but the sampling process itself can be disruptive to the research program. Since every plant is needed to maximize return of seed for next spring planting, the removal of just one kernel results in a 5% stand reduction. The adverse impact that this procedure can have on the overall program serves to highlight the need for regulatory agencies throughout the world to adapt their sampling procedures to accommodate winter nursery activities.

Figure 2. Winter nursery in Hawaii. (Courtesy of Pioneer Hi-Bred International Inc.)

Post-entry grow-outs

Quarantine isolation fields are often set up as a condition for seed entry. Companies must check to ensure that seed sources sent to the nurseries have been certified as free of the quarantine pests and pathogens specified by the country. The post entry isolation fields are carefully monitored by government and company pathologists for any indication of seed-borne pathogens, and if any quarantined pests are detected, the plants are immediately destroyed and appropriate eradication measures enacted.

Security of Proprietary Breeding Material

Companies often require that special precautions be taken by government agencies to safeguard misappropriation of their proprietary breeding material. Each company has their own guidelines to ensure germplasm security, and custom and phytosanitary inspectors are often not aware of these concerns when requesting seed samples to test for seed-borne pathogens. Agreements then have to be reached to meet the needs of both parties.

SEEDBORNE DISEASE DATABASES

Royer and Dowler (1988) described the need for a worldwide database for plant pathogens to provide accurate information on the host range, distribution, and favorable environmental conditions for plant pathogens. CAB International is in the process of constructing an Electronic Crop Protection Compendium that will comprise these and many other features for diseases of all major crops of the world. A database on seed-borne diseases will also be included in the product. Computer databases also have been established by many state plant disease diagnostic clinics, government disease surveying agencies, and by private companies for internal use or for use by customers in a state. National data bases such as that proposed some years ago by the Plant Protection and Quarantine (PPQ) unit of the United States Department of Agriculture (USDA) and the Intersociety Consortium for Plant Protection (ICPP) would be of benefit in establishing a centralized system to collect, summarize, and distribute information on major agricultural pests (Wallenmaier, 1986). When used with complex meteorological models, databases of this nature can be used to predict risk probabilities if an exotic pest is introduced into a country.

For example, the CLIMEX model is currently used to generate risk indices for entomological agricultural pests of Australia (Sutherst and Maywald, 1988). Yang *et al* (1991) used the introduction of soybean rust into the US as an example of how the impact of a disease on a previously uncontaminated area may be predicted by its ecological requirements in the geographic areas where the pathogen and host are already distributed. Such systems can be very useful to agricultural scientists and quarantine officials in determining the real risk potential for accidental pest introduction at any site (Worner, 1988) including winter nursery locations.

Databases that keep track of regulations impacting movement of seeds are of great value to commercial and private plant breeding programs. One such system is the USDA-APHIS EXCERPT network which allows seed companies to access phytosanitary regulations for most countries throughout the world. The regulations are continually updated to reflect any changes. The CAB International seed-borne disease database also will provide records on disease incidence and pathogen distribution that can be used to review phytosanitary regulations that restrict seed movement. Since many of these regulations were put into effect many years ago, they need to be reevaluated in the light of new scientific evidence.

Cooperative efforts between industry and government agencies on seed increase and varietal improvement have been established over the past few years. The Latin American Maize Program (LAMP) coordinates the efforts of plant breeders to characterize maize germplasm collections from ten countries and often includes references to disease resistance. The information is loaded into a common database and made available to public and private company plant breeders (Smith *et al*, 1989). Similar programs that focus on diseases are needed between private and government plant pathologists. Resources also need to be devoted to establishing well-defined, uniform testing methods for detecting seed-borne pathogens.

CONCLUSION

Multinational seed companies recognize the need to protect against the spread of plant pathogens by seeds between production locations both for their own commercial interests and for world food production in general. The worldwide phytosanitary system is a necessary regulatory mechanism to ensure this. Combined efforts of industry, regulatory, and university scientists are needed to provide improved information sources, however. Improved communication between industry and regulatory agencies, standardized seed health testing measures, together with the careful inspection of nurseries by government and commercial pathologists, will not only reduce the risk of accidental introduction of exotic plant pathogens by means of seed, but also encourage efficient and expeditious international movement of germplasm.

LITERATURE CITED

Duvick, D. N. 1991. Possible genetic causes in increased variability in U.S. maize yields. Pages 492-498. in: *Ethics and Agriculture. An Anthology on Current Issues in World Context.* ed. C. V. Blatz, Univ. of Idaho Press, Moscow. 674 pp.

McGee, D. C. 1988. *Maize Diseases: a Reference Source for Seed Technologists.* APS Press, St. Paul. 150 pp.

Royer, M. H. and Dowler, W. M. 1988. A world plant pathogen database. Plant Dis. 72: 284-288.

Smith, J. S. C., and Duvick, D. N. 1989. Germplasm collections an the private plant breeder. Pages 17-30 in*: The Use of Plant Genetic Resources.* eds. A. H. D. Brown., O. H. Frankel, D. R. Marshall, and J. T. Williams. Cambridge Univ. Press., Cambridge. 382 pp.

Sutherst, R. W., and Maywald, G. F. 1985. A computerized system for matching climates in ecology. Agric. Ecosystems Environ. 13: 281-299.

Thurston, H. D. 1973. Threatening plant diseases. Annu. Rev. Phytopathol. 11: 27-52.

Wallenmaier, T. E. 1986. Development of the computerized state-federal pest reporting system. Plant Dis. 70:365-367.

Worner, S. P. 1988. Ecoclimatic assessment of potential establishment of exotic pests. J. Econ. Entomol. 81: 973-983.

Yang, X. B., Dowler, W. M., and Royer, M. H. 1991. Assessing the risk and potential impact of an exotic plant disease. Plant Dis. 75: 976-982.

CHAPTER 3

ACTIVITIES AT THE INTERNATIONAL AGRICULTURAL RESEARCH CENTERS TO CONTROL PATHOGENS IN GERMPLASM

M. Diekmann

Germplasm Health Scientist
International Plant Genetic Resources Institute,
142 Via delle Sette Chiese, 00145 Rome, Italy

The Consultative Group on International Agricultural Research (CGIAR) was established jointly in 1971 by the World Bank, the United Nations Development Programme (UNDP), and the Food and Agriculture Organization of the United Nations (FAO). It is a broadly based consortium of 41 public and private sector donors supporting a network of 17 international agricultural research centers (IARCs) located throughout the world, mainly in developing countries. The centers of the CGIAR maintain, in trust for the world community, large

Figure 1. Germplasm storage facility (Courtesy of Iowa State University).

collections of genetic resources of major food and forage crops. In total, more than 500,000 accessions are held in 11 different genebanks within the CGIAR system (Fig. 1). This material is freely available to bona fide users and is exchanged world-wide. Many centers are breeding new varieties which have resistance to adverse environmental conditions as well as to pests and diseases, and produce good yield with low input. These newly developed lines need to be tested under different conditions, which involves distribution of test sets to other countries and regions.

IARC GERMPLASM EXCHANGE ACTIVITIES

Germplasm is handled by thirteen IARCs (Table 1). All of these centers engage in extensive germplasm exchange. The IARC germplasm collections form the largest international repository of genetic diversity of the major food crops of the world. About 95% of the germplasm is conserved and exchanged in the form of true seeds; only 5% is vegetatively propagated or kept and exchanged in the form of tissue culture (e.g., potato, cassava, *Musa* spp.). Between 1987 and 1991 approximately 750,000 samples of germplasm proper were distributed, 250,000 of them in the host country, and 350,000 to another IARC. In general, the number of samples distributed exceeds that of samples received; in the case of CIAT, for example, by almost 100% (Cooper *et al*, 1994). In addition to germplasm in the strict sense (i.e., the material that is stored in genebanks) a vast amount of 'breeding lines' (e.g., segregating populations and other unfinished material) is exchanged. Since 1974, ICRISAT has dispatched almost one million seed samples to 153 countries. In 1992, IRRI sent almost 100,000 samples in 515 shipments to 67 countries, most of them as part of the International Network for Genetic Evaluation of Rice (IRRI, 1992).

When measured in terms of total weight, germplasm material is relatively small compared with commercial seed shipments however, the diversity of germplasm poses special problems. For example, different lines can be expected to have different levels of resistance to pests and pathogens. Another problem is that germplasm collections usually contain seeds from regions of the world that are not only centers of origin of the crop, but may also be centers for genetic diversity of crop-specific pathogens (Leppik, 1970). Seeds from these regions, therefore, may be contaminated by pathogenic strains or races exotic to the final location. This problem is exemplified by the severe

Table 1. Location and crop responsibilities of International Agricultural Research Centers handling plant germplasm

Center	Crops
CIAT Cali, Colombia	Cassava, rice, beans (*Phaseolus*), tropical forages
CIFOR Bogor, Indonesia	Forest tree species
CIMMYT Mexico 06600 D.F., Mexico	Maize, wheat
CIP Lima, Peru	Potato, sweet potato, Andean root and tuber crops
ICARDA Aleppo, Syria	Cereals
ICRAF Nairobi, Kenya	Multipurpose trees
ICRISAT Patancheru, India	Sorghum, millet, groundnut, chickpea, pigeonpea
IITA Ibadan, Nigeria	Maize, rice, grain, legumes, banana, plantain, root and tuber crops
ILRI Addis Ababa, Ethiopia	Forages
INIBAP/IPGRI Montferrier-sur-Lez, France	Banana, plantain
IPGRI Rome, Italy	Not crop specific
IRRI Manila, Philippines	Rice
WARDA Bouaké, Côte d'Ivoire	Rice

economic consequences for commercial pea production in North America resulting from the introduction of pea seed-borne mosaic virus into pea breeding lines from a germplasm collection (Hampton *et al*, 1993).

PATHOLOGICAL CONCERNS IN GERMPLASM

While this monograph addresses the movement of pathogens by seeds, plant pathogens also impact germplasm in other ways. One is the effect on seed longevity. There is evidence that infected seeds may lose their viability sooner than healthy ones. For example, storage fungi (*Penicillum* and *Aspergillus* spp.) can negatively impact germinability of all types of seeds, particularly if they stored under high humidity (Christensen and Meronuck, 1986). This can mean expensive regeneration, which also entails the risk of losing some lines due to adverse conditions. Plant diseases can also influence germplasm characterization and evaluation. For example, several viruses and some fungi such as *Colletotrichum gloeosporioides, Pyricularia oryzae,* or *Sclerospora graminicola* may obscure key morphological characteristics needed to classify plants. A final concern is to ensure that genetic diversity of the germplasm collection is maintained when pathogen elimination programs are implemented (Klein *et al*, 1990).

The most serious impact of pathogen infection/contamination on germplasm, however, is the international movement of these pathogens. The IARCs have become particularly concerned about this in the past 10 years and carry out a number of activities to control the spread of pathogens.

IARC ACTIVITIES TO CONTROL PATHOGENS IN GERMPLASM

Direct Control

Most centers have established seed health/germplasm health laboratories with various staffing levels. These units report to the Director General (CIMMYT), the Deputy Director General (IRRI), or the Head of the Genetic Resources Unit (CIAT, ICARDA, ILRI). Almost all centers have a germplasm health or seed health committee that comprises crop protection specialists, virologists, and plant pathologists. ICRISAT and IRRI also have representation from the host country's plant quarantine service. At CIAT, ICRISAT and IRRI, phytosanitary certificates are issued by a quarantine officer assigned to the institute. CIAT, CIMMYT and CIP issue Germplasm Health Statements which complement the phytosanitary certificates from the host country by providing additional information on the test methods or therapy applied.

CIAT, CIP, ICARDA, IITA and IRRI have both containment facilities and isolated fields which are used for post-entry quarantine. CIAT, IITA, ILRI, and WARDA regenerate their germplasm material in isolation to some extent. A systematic sanitation of the collection is routinely done by CIAT for cassava, by CIP for potato and sweet potato, and by INIBAP for *Musa* banana bunchy top virus (BBTV) and cucumber mosaic virus (CMV). In addition, IITA has plans to free the cowpea collection of seed-borne pathogens and CIAT intends to do the same for *Phaseolus* beans.

ICARDA has applied the following seed health protocol (Diekmann, 1988) and other centers have adopted similar systems suiting their particular needs:

- *Seed regeneration in areas with low disease pressure.*
- *Field inspection by experienced pathologists/virologists.*
- *Pesticide application if appropriate.*
- *Seed health testing.*
- *Seed treatment.*

Seed Health Research Activities

Procedures required for a safe and fast exchange of germplasm are often not available. Many of the IARCs carry out research projects aimed at solving such problems which may also be carried out jointly with universities or other research organizations.

ICARDA is studying the effect of dry heat applied to lentil seeds on virus eradication. Heat treatment of 70°C eliminated seed-borne infection with broad bean stain virus (BBSV) although it reduced germination to 40% (K. M. Makkouk, *pers. com.*). Further studies are planned by ICARDA for material infected with pea seed-borne mosaic virus (PSbMV) and bean yellow mosaic virus (BYMV). Since the dry heat method can also be expected to control bacteria and fungi, it may be used in rescuing valuable infected germplasm.

IRRI has investigated control of rice blast with antagonistic bacteria. From 400 bacterial strains tested in the laboratory, two strains of *Pseudomonas fluorescens* and two of *Bacillus* spp. gave good inhibition of *Pyricularia oryzae*. Seeds were coated with the bacterial suspension and disease development was compared with that in fungicide-treated seeds. Bacterial treatments gave significant disease control, but were not as effective as the fungicide (Gnanamanickam and Mew, 1992). IRRI has also been active in the area of detection of seed-borne infection. Primers were selected and tested against genomic DNA from different strains of *Xanthomonas* species and other bacteria. Three primer pairs could distinguish between the pathovars *Xanthomonas oryzae* pvs. *oryzae* and *oryzicola*

Figure 2. Kernel infection by *Tilletia indica* (Karnal bunt) in wheat seeds (Courtesy of CIMMYT).

and could be used for the detection of *X. oryzae* pv. *oryzae* from artificially infected leaves and seeds (IRRI, 1992).

CIP has been involved for many years in the refinement of methods for virus detection in potato propagation material including botanical seeds. Potato virus T (PVT) and potato spindle tuber viroid (PSTVd) are the most important agents known to be transmitted in true seed. A nucleic acid spot hybridization (NASH) test with probes for the detection of both pathogens was developed to test for the pathogens in seed and as a consequence, only progeny of the pathogen-tested parents are distributed (CIP, 1992).

CIMMYT has focused its efforts on seed-borne diseases around Karnal bunt of wheat (Fig. 2). The center spent almost one million US$ on Karnal bunt research between 1982 and 1986 (Klatt, 1986). In addition to seed multiplication in areas free from Karnal bunt and foliar application of the fungicide propiconazole, a system was

Figure 3. Core of corn tissue removed for a nondestructive seed health assay (Courtesy of Iowa State University).

devised to wash all internationally distributed wheat seed in 1% sodium hypochlorite to destroy seed-borne teliospores of *Tilletia indica* (Butler and Muhtar, 1993).

IPGRI's research projects are always carried out in collaboration with other research institutions. In the past, the germplasm health projects covered seeds as well as vegetatively propagated crops. One example is the development of non-destructive seed health testing carried out in collaboration with the Iowa State University Seed Science Center (Fig. 3). The results of this project showed that there are ways to test and save the seeds of large-seeded crops (Higley *et al*, 1993). Another seed-related project was started in 1994 in cooperation with the Asian Vegetable Research and Development Center on detection of *Xanthomonas campestris* pv. *vesicatoria* in seeds of tomato and pepper.

Seven centers have recently started a collaborative research project to study the effect of pathogens on seed longevity that will cover viruses, bacteria, and fungi in different crops.

Training

Some centers have conducted specialized courses in topics related to seed pathology. Courses at CIP, ICARDA, and IRRI concentrated on seed production or the production of healthy planting material. In 1992, INIBAP organized a regional quarantine training course for banana and plantain in Burundi and trained two scientists from Venezuela in quarantine principles for black Sigatoka diseases. CIAT trained quarantine officers in the detection of seed-transmitted viruses of beans. CIMMYT, in collaboration with the Iowa State University Seed Science Center, carried out seed health training for Mexican plant health officials. ICARDA provided individual training for three quarantine officers and equipped a seed health laboratory for the Department of Agriculture in Aleppo.

Information Exchange

A number of centers have recently published the following books on seed health.

- *IRRI (A Manual of Rice Seed Health Testing).*
- *ICRISAT (A Pictorial Guide to the Identification of Seedborne Fungi of Sorghum, Pearl Millet, Finger Millet, Chickpea, Pigeonpea, and Groundnut).*
- *ICARDA (Seed Health in Seed Production, Seed-borne Pests and Diseases of Faba Bean).*

In 1988, IPGRI began jointly publishing with FAO, a series of Technical Guidelines for the Safe Movement of Germplasm and to date, 17 guidelines have been published.

CONCLUSION

The heightened awareness in recent years of concerns regarding the dissemination of plant pathogens by seeds, together with a proliferation of new phytosanitary regulations, has forced International Agricultural Research Centers (IARCs) in the CGIAR system to take action to protect against inadvertent introduction of plant pathogens by germplasm exchange. The quantity and diversity of the genetic material exchanged throughout the world makes this a much more complicated technical and regulatory issue than that for shipments of commercial seeds. Implementation of adequate controls of pathogen movement in germplasm will remain a continuing challenge for the CGIAR system, particularly in the face of dwindling economic resources.

LITERATURE CITED

Butler, L., and Muhtar, H. 1993. Elimination of teliospores of *Tilletia indica* Mitra on wheat seed surfaces by mechanized disinfestation with sodium hypochlorite. 6th International Congress of Plant Pathology, July 28 to August 6, 1993 in Montreal. Abstract No. 15.1.20, p. 262.

Christensen, C. M., and Meronuck, R. A. 1986. *Quality Maintenance in Stored Grains and Seeds.* University of Minnesota Press, Minneapolis, MN. 138 pp.

CIP, 1992. Annual Report 1992. International Potato Center, Lima, Peru. 222 pp.

Cooper, D., Engels, J., and Frison, E. 1994. A multilateral system for plant genetic resources: imperatives, achievements and challenges. Issues in Genetic Resources No. 2, May 1994. International Plant Genetic Resources Institute, Rome, Italy.

Diekmann, M. 1988. Seed health testing and treatment of germplasm at the International Center for

Agricultural Research in the Dry Areas (ICARDA). Seed Sci. Technol. 16:405-416.

Gnanamanickam, S. S., and Mew, T. W. 1992. Biological control of blast disease of rice (*Oryza sativa* L.) with antagonistic bacteria and its mediation by a *Pseudomonas* antibiotic. Ann. Phytopath. Soc. Japan 58:380-385.

Hampton, R. O, Kraft, J. M, and Muehlbauer, F. J. 1993. Minimizing the threat of seedborne pathogens in crop germ plasm: elimination of pea seedborne mosaic virus from the USDA-ARS germ plasm collection of *Pisum sativum*. Plant Dis. 77:220-224.

Higley, P. M., McGee, D. C., and Burris, J. S. 1993. Development of methodology for non-destructive assay of bacteria, fungi and viruses in seeds of large-seeded field crops. Seed Sci. Technol. 21:399-409.

IRRI. 1992. Program Report for 1992. International Rice Research Institute, Los Baños, Philippines. 316 pp.

Klatt, A. 1986. Summary and comments. pp 34-36 In: *Proceedings of the Fifth Biennial Smut Workers' Workshop, April 28-30, 1986.* CIMMYT, Mexico City, Mexico. 37 pp.

Klein, R. E, Wyatt, S. D., and Kaiser, W. J. 1990. Effect of diseased plant elimination on genetic diversity and bean common mosaic virus incidence in *Phaseolus vulgaris* germ plasm collections. Plant Dis. 74:911-913.

Leppik, E. E. 1970. Gene centers of plants as sources of disease resistance. Annu. Rev. Phytopathol. 8:323-344.

ACKNOWLEDGEMENTS

I would like to thank my colleagues at other International Research Centers for supplying information for this paper: Dr. Francisco Morales, CIAT; Dr. Larry Butler and Dr. Jesse Dubin, CIMMYT; Dr. Luis Salazar and Dr. Ed French, CIP; Dr. Khaled Makkouk and Dr. Ahmed El Ahmed, ICARDA; Dr. A.K. Ghanekar, ICRISAT; Dr. Diane Florini and Dr. Annemiek Schilder, IITA; Dr. Jean Hanson, ILRI; Dr. David Jones, INIBAP/IPGRI; Dr. Tom Mew, IRRI; Dr. Abdoul Aziz Sy, WARDA. Thanks are also due to Ms. Linda Sears, IPGRI for editing the manuscript.

CHAPTER 4

SEED QUARANTINE SYSTEMS IN THE UNITED STATES

N. G. Klag

Operations officer
USDA, Animal and Plant Health Inspection Service,
Plant Protection and Quarantine
4700, River Rd, Riverdale, MD 20737

The import and export of seed is very important to commerce and to scientific exchange of germplasm. This paper will discuss the role of the United States government and specifically that of the US Department of Agriculture Animal and Plant Health Inspection Service (USDA-APHIS) in preventing the international spread of plant pathogens by seeds.

STRUCTURE OF USDA-APHIS

USDA-APHIS is divided into ten units.

- *Animal Damage Control.*
- *Biotechnology, Biologics, and Environmental Protection.*
- *International Services.*
- *Legislative and Public Affairs.*
- *Management and Budget.*
- *Human Resources Division.*
- *Plant Protection and Quarantine.*
- *Recruitment and Development.*
- *Regulatory Enforcement and Animal Care.*
- *Veterinary Services.*

Plant Protection and Quarantine (PPQ) is the unit that regulates the export and import of seeds. PPQ has three basic missions:

- *To protect the nation's agricultural resources from exotic plant pests and diseases.*
- *To prevent the spread of exotic plant pests within the United States.*
- *To certify US agricultural products for export.*

This paper will address the first and third points, namely import and export of agricultural products and will focus specifically on seed.

ENABLING LEGISLATION FOR SEEDS IMPORTED INTO THE US

A series of legislative actions since the early part of the 20th century controls the regulations pertaining to importation of seeds into the US.

Plant Quarantine Act of 1912

Prior to the passage of this act, the USDA had no control over the importation of plants and plant products or their movement from Hawaii, Puerto Rico or the Virgin Islands to the mainland. This act and the regulations issued under its authority permits PPQ to restrict or prohibit the entry of host plants (including their seed) in order to protect US agriculture from specific plant pests and pathogens.

Three examples of quarantine regulations promulgated under this act include:

- *The Bamboo Quarantine (7CFR 319.34) prohibits entry of bamboo seed capable of propagation from all foreign countries except into Guam.*
- *The Nursery Stock Quarantine (7CFR 319.37) restricts entry from foreign countries of a number of seed genera by requiring that such seed be treated prior to entry.*
- *The Corn Diseases Quarantine (7CFR 319.24 and 7CFR 319.41) prohibits the entry of corn seed and many related plants from certain countries because of diseases and injurious insects. The diseases mentioned in the corn regulation (7CFR319.24) are Physoderma zeae-maydis, Physoderma maydis, Peronospora maydis, Sclerospora sacchari and other downy mildews.*

Federal Seed Act of 1939

This act restricts the entry of seed, described as "Agricultural" or "Vegetable", to ensure seed purity: namely that the seed is what its label says and that the seed lot is free from noxious weeds as identified in the Federal Seed Act. As a note of interest, except for dodder (*Cuscuta* spp.), the noxious weeds listed under the FSA are not the same as those listed as noxious under the Federal Noxious Weed Act. The nine weeds listed under the FSA are already established and widespread in the United States but are serious problems when introduced into a field with crop seed. A small tolerance is allowed for these weeds, in that three or more weed seeds in a specified amount of crop seed must be found for entry to be refused. If an agricultural or vegetable seed shipment is refused for noxious weed content, the consignee is allowed to clean the seed under PPQ supervision.

Federal Plant Pest Act of 1957

This act and the regulations issued under its authority gives authority to inspectors to take emergency action to seize, treat, or destroy articles or products with respect to plant pests that are new or not widely prevalent in the United States. The Act defines "plant pest" as "any living stage of ....parasitic plants or reproductive parts thereof...". Therefore, seeds of parasitic weeds new to or not widely prevalent or distributed within and throughout the United States are prohibited. This prohibition covers seeds of parasitic plants imported as such and extends to those seeds when found in imported seed shipments.

Endangered Species Act of 1973

This act and the regulations promulgated under its authority provides for the protection of listed species by prohibiting the unauthorized importation of these plants and their seed.

Federal Noxious Weed Act of 1974

This act and the regulations issued under its authority restricts the entry of weeds and their seeds which are determined to be harmful to agricultural crops, livestock, irrigation, navigation, fish and wildlife resources, or public health.

Convention on International Trade in Endangered Species of Fauna and Flora

This is a multinational treaty that limits the import, export, and re-export of listed species of wild flora and fauna.

Most seed is allowed entry into the United States based on visual inspection at the port of entry. If a pest of quarantine significance is detected, the seed is either rejected or a seed treatment is prescribed, if available. A number of seed species are prohibited or can only enter under a permit which prescribes the conditions of entry, as described (USDA/APHIS, 1993).

PLANT QUARANTINE REQUIREMENTS FOR SEED EXPORTS

Role of APHIS-PPQ

APHIS-PPQ serves to assist exporters in meeting the plant quarantine requirements of foreign countries. The role of PPQ in facilitating seed exports is as follows:

- *Maintain current information on plant quarantine import requirements of foreign countries.*
- *Analyze information, prepare summaries of import requirements based on the information, and distribute information to federal, state, and county certifying officials, shippers, nurseries, and other interested parties. This information is accessible from the computer data base system, EXCERPT, and most countries to which the United States exports agricultural commodities are now in the system. The system is also available to all federal and state regulatory personnel. Public access is available on a fee basis.*

- *Inspect domestic plants and plant products to ensure that they that meet the import requirements of the foreign country.*
- *Monitor the issuance of certificates to ensure their accuracy.*
- *When possible, assist US exporters if their shipments are held at the destination.*
- *Assist in certifying plants or plant products of foreign origin that have been legally imported into the US and are being re-exported.*
- *Ensure that only authorized federal, state, and county officials inspect and certify the phytosanitary status of plants and plant products offered for export.*
- *Prevent the detention or destruction of US agricultural commodities in receiving countries.*

The primary means of certification is by the issuance of Federal Phytosanitary Certificates (FPC). The phytosanitary certificate has standard uses, rules, and format, which were determined by the International Plant Protection Convention of 1951 and is recognized and used by most countries of the world. Under this treaty, APHIS has the responsibility to ensure that US seed exports meet the standards of the international phytosanitary certificate and the phytosanitary requirements of the importing country.

Role of State Agencies

APHIS-PPQ has approximately 1,300 employees, of which 850 are inspectors located at every major seaport and airport and in most states. Because of increasing demands for certification, a cooperative program was set up in 1975 to authorize state regulatory inspectors to issue FPC's. This is accomplished by a signed agreement between the Deputy Administrator of PPQ and the primary state regulatory official. The agreements define the eligibility requirements for inspectors and their obligations. State inspectors are trained by PPQ and a list of eligible inspectors is maintained. Oversight of this program is by PPQ, and there are currently about 1,200 approved State inspectors.

Federal Phytosanitary Certificate (FPC)

Export certification is a service provided to the seed industry. Obtaining an FPC is not a requirement of the USDA to export plant products. It is the importing country that gives the initial and final authority on the entrance of the plant material. If a country declares a product prohibited, PPQ cannot issue an FPC without a special import permit from that country.

An FPC states the following:

- *The country of origin.*
- *That the plants or plant products are considered to be free from quarantine pests.*
- *That the plants or plant products are practically free from other injurious pests.*
- *That the plants and plant products meet the entry requirements of the importing country.*

Foreign seed regulations often require a declaration that the product is free from specific pests. This can be done, usually at the option of the importing country, by:

- *Visual inspection of the product.*
- *Inspection of the mother plants during growth in the field.*
- *Laboratory testing of the seed. Currently laboratory testing can only be conducted at federal, state, or university seed testing laboratories.*

Figure 1. Stunted corn plants infected by *Peronosclerospora sorghi* (Courtesy of CIMMYT).

Phytosanitary requirements can vary between countries for specific commodities. Examples for corn seed are as follows:

Taiwan - An FPC is required. Specific requirements are not known. It is recommended that the exporter request an import permit.

Panama - An FPC and import permit are required. The parent plants must be inspected during active growth and found to be free of sorghum downy mildew caused by *Peronosclerospora sorghi*, (Fig. 1) or the parent plants must be located in areas known to be free from sorghum downy mildew.

An additional declaration is needed on the FPC attesting to one of these conditions.

Figure 2. Corn seedling infected with *Clavibacter michiganense* subsp. *nebraskense* in a grow-out test (Courtesy of Iowa State University).

Mexico - The seed must be free of *Erwinia stewartii* and *Corynebacterium michiganensis* subsp. *nebraskense* (Fig. 2). Freedom from these pests can be determined by field inspection of the mother plants in active growth or from laboratory analysis of an officially drawn, representative sample of seed.

The seed also must not have more than a 2% infection rate with *Sclerophthora macrospora.* This tolerance must be determined by a laboratory analysis of an official sample of seed. If the seed is determined to be free of this disease based on an active growth field inspection of the mother plants, then the established tolerance is met.

An additional declaration is required stating that "The seed in this shipment is free of *Erwinia stewartii* and *Corynebacterium nebraskense*. The seed meets the 2% tolerance for infection with *Sclerophthora macrospora*."

Role of the Seed Industry

Export certification begins with the production of high quality seed by the grower. The exporter must then present a product that is able to meet the entry requirements of the importing country which often requires appropriate cleaning or culling. The product must also be properly inspected to ensure that it is of the highest quality. The USDA officer or state official then does the final inspection and the product is ready for export at this point. All parties must work together in order to provide a product that meets the phytosanitary requirements of the importing country.

NEW US SEED HEALTH ACCREDITATION SYSTEM (USSHI)

The US has managed to maintain the respect of the international plant protection community by managing a certification program that in 1995 issued over 270,000 phytosanitary certificates while receiving only 13 valid complaints. US agricultural exports were expected to total over $50 billion in 1995. Of this total, it was estimated that approximately $23 billion worth of products would be accompanied to destination by Federal Phytosanitary Certificates.

With the growing export market for US seeds and a concomitant increases in phytosanitary regulations, the American Seed Trade Association submitted a proposal in

1995 to accredit seed company and independent laboratories to carry out laboratory seed health test and field inspections that would be used to obtain phytosanitary certificates. This proposal is now undergoing the process of Federal approval. This program would accelerate the pre-export phytosanitary process while ensuring the accuracy and consistency of analysis in field inspections and seed health testing by non-governmental entities.

This proposal is a part of trend toward governments loosening control and giving the responsibility to the seed industry to regulate itself (USDA/APHIS, 1994). This process is already well-underway in Canada and the Netherlands (Wesseling, 1997).

CONCLUSION

The integrity of individual seed health testing laboratories will be greatly enhanced by the standards set for facilities, equipment, personnel, and health tests in the new seed health accreditation system, as also will be the quality of performance and international reputation of USDA-APHIS, as the government agency responsible for implementing the international phytosanitary certification system.

LITERATURE CITED

USDA/APHIS 1993. Entry status of seeds for planting under federal plant quarantines. Circ. PPQ Q.37-4 04/93.

USDA/APHIS 1994. *"The Seed Industry: A Background paper"* Policy and Analysis Department, United States Department of Agriculture, Animal and Plant Health Inspection Service, October, 1994.

Wesseling, J. B. M. 1997. Government/Vegetable Seed Industry cooperation in the Netherlands. Pages 93-103 in: *Plant Pathogens and the Worldwide Movement of Seeds,* ed. D. C. McGee. APS Press, St. Paul, MN. 109 pp.

CHAPTER 5

WORLD PHYTOSANITARY SYSTEM: PROBLEMS AND SOLUTIONS

D. C. McGee

Professor of Plant Pathology
Seed Science Center, Iowa State University
Ames, IA 50011

Seeds can be efficient means of moving pathogens between geographical regions. There are many documented examples of introductions of pathogens from one country to another by seeds that resulted in major economic losses (Neergaard, 1977). Regulations clearly are necessary, therefore, to protect against the international spread of plant pathogens. Seed lots that move between countries must be accompanied by a standard phytosanitary certificate. This was defined by the International Plant Protection Convention of 1951 and states that seeds are substantially free of injurious pests or diseases. The certificate also includes a section to deal with specific diseases or pests for which the importing country may require tests. In addition, a country may have specific requirements regarding types of seeds and pathogens that may be admitted. Examples include:

- *Importation of certain seeds may be prohibited.*
- *Seeds are re-tested for pathogens after entry into the importing country.*
- *Seed must be treated with a fungicide.*

The seed health information for the phytosanitary certificate is generated and reported by the country exporting the seed. This information may be determined either by a field inspection of the growing crop or by a laboratory assay of harvested seeds.

It is commonly accepted by industry, research, and regulatory institutions involved in the international movement of seeds that the current world phytosanitary system has serious weaknesses. This paper will discuss the reasons for and consequences of the problems in the system, and will suggest possible solutions.

PROBLEMS WITH THE WORLD PHYTOSANITARY SYSTEM

Current problems with the world phytosanitary system can be related to the following factors:

Poor Understanding of Potential Economic Losses from Introducing Pathogens

This problem can be attributed to limited access to information on the pathogen and to insufficient research into the epidemiology of seed-borne diseases. A consequence of the problem is the proliferation of

Figure 1. Soybean leaves infected by *Pseudomonas syringae* pv. *glycinea* (bacterial blight) (Courtesy of Iowa State University).

phytosanitary regulations for pathogens that cannot be justified with respect to the economic risk posed by their introduction into the importing country. For example, all soybean seed lots exported from the US to the European Union (EU) are required to be tested for *Pseudomonas syringae* pv. *glycinea*, the cause of soybean bacterial blight (Fig. 1). This disease has been present in Europe for many years (Laszlo, 1982). It is also widespread in the US but has minimal economic impact (Wilcox, 1983, McGee, 1992). Since 1988, US seed companies have spent approximately $1,500,000 on testing for this pathogen on soybean seeds to meet this phytosanitary requirement. The need to meet requirements for unjustified regulations not only impairs international movement of seeds worldwide, but wastes resources that could be applied to protecting against the spread of economically important seedborne pathogens that often are ignored. *Diaporthe phaselorum* var. *meridionales*, the cause of southern stem canker of soybean (Fig. 2),

Figure 2. Soybean cultivars susceptible (left) and resistant (right) to *Diaporthe phaselorum* var. *meridionales* (Courtesy of W. F. Moore).

is a seed-borne pathogen that falls into this category. This is a relatively new disease that was first detected in Mississippi in 1977, and by 1984 was widespread throughout the southeast US, causing massive crop losses (Backman *et al*, 1985). Since 1989, southern stem canker has become a major problem in Brazil and Paraguay (Yorinori, 1990, Sato *et al*, 1993). There is no evidence that the pathogen was introduced by US soybean seeds, but Brazil had no phytosanitary regulation to test for this pathogen in soybean seed imports from the US at that time.

Lack of Knowledge of Relationships Between Tolerances in Seed Assays to Risks of Transmission of the Pathogen to the Planted Crop

This problem is also a result of lack of access to information on the pathogen and of insufficient research into the epidemiology of seed-borne diseases. One consequence of the problem is that it leads to the use of non-scientifically based tolerances by default. Again *Pseudomonas syringae* pv. *glycinea* provides an example. This pathogen must be detected in each of five sub-samples of 1 kg of soybean seeds for the seed lot to be denied a phytosanitary certificate. This tolerance was determined by negotiation between US and EU authorities and has no scientific basis as a means of protecting against transmission of the pathogen by seeds (Author, *pers. com.*).

A second consequence of this problem is that sample sizes may be inadequate or impractical. For example, a grow-out test, commonly used to detect *Erwinia stewartii* in corn seeds, requires 400 seeds (McGee, 1988). In the 15 years that the test has been employed in the Iowa State University Seed Testing Laboratory, the pathogen has never been detected in a corn seed lot. Recent research indicates that 10,000 seeds would have to be tested to give a realistic chance of detection of *Erwinia stewartii* (Block, 1996). The space and labor required to test 10,000 corn seeds would be impractical for both logistical and economic reasons.

Lack of Standardization in Testing Protocols

Apart from 68 working sheets on seed-borne diseases produced by the International Seed Testing Association, Zurich, Switzerland, there has been

Figure 3. Grow-out test for *Erwinia stewartii* infection of corn seeds (Courtesy of Iowa State University).

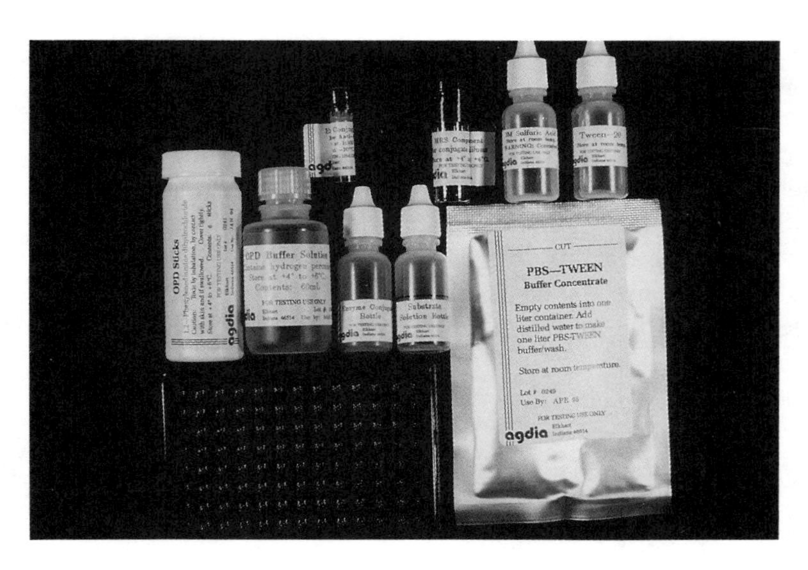

Figure 4. ELISA kit for detection of *Erwinia stewartii* infection of corn seeds (Courtesy of Agdia Inc.).

no systematic effort to develop standardized seed health test methods that are accepted internationally. A consequence of this is that different tests may be used by exporting and importing countries. In 1994, *Erwinia stewartii* was detected by a serological procedure (Fig. 4) in the Netherlands in a shipment of corn seeds that had been issued a phytosanitary certificate in the US based on a grow-out test (Fig. 3). The shipment of 20 tonnes of hybrid seed had to be returned to the US (Author, *pers. com.*). Experiences such as this usually remain confidential, but seed industry personnel assert that they occur with considerable frequency (Author, *pers. com.*).

A second consequence of this problem is that there are numerous seed health tests in common use that have never been subjected to international standardization. The seed health testing program at Iowa State University, for example, offers approximately 70 tests, only five which are published as working sheets of ISTA and could thus be considered as standardized (Author, *pers. com.*).

Different Seed Treatment Regulations Among Countries

Regulations and standards for efficacy and environmental risk of pesticides vary considerably between countries. For example, most countries now ban organo-mercury products. These have been replaced by a range of new materials, such as the ergosterol biosynthesis inhibitors, some of which are available as seed treatments in Europe but not in the US (McGee, 1996). The main consequence of this problem is confusion regarding the legality of use by the exporting country of particular treatments and their rates of application.

Trade Barriers

Trade barriers result from the use of phytosanitary certification to restrict importation of seeds for political/economic reasons. There is strong evidence that trade barriers are becoming more prevalent with the development of free trade areas such as NAFTA and the EU, and by completion of the recent GATT agreement. Because these agreements remove tariffs as barriers to trade, other instruments such as phytosanitary certification have the potential to be used to achieve the same purpose.

SOLUTIONS TO PHYTOSANITARY PROBLEMS

The following elements are needed to provide a better international phytosanitary system for seed-borne pathogens: improved information sources, standardized test methods, improved access to seed treatment regulations, and revised seed pathology research priorities.

Improved Information Sources for Seed-Borne Pathogen

A good starting point to obtain information on seed-borne diseases is the Annotated List of Seed-borne Diseases (Richardson, 1990). This publication lists all seed-borne microorganisms that have been recorded in the scientific literature as being associated with seeds of all crops. Unfortunately, as previously stated, the publication can be misused because it does not provide adequate information to justify phytosanitary regulations such as the economic risk from introducing the pathogen or whether the pathogen can be transmitted by seeds. In 1994, the

American Seed Trade Association commissioned analyses of the economic risks of transmission of cherry leaf roll virus by soybean seeds and *Periconia circinata* by sorghum seeds. These pathogens had been proposed as phytosanitary regulations against US seeds. Results showed that soybeans are not a natural host for cherry leaf roll virus and that *P. circinata* had not been found in the US for 40 years, nor is it recorded as transmitted by sorghum seeds. Authorities in the importing countries, however, cited the presence of these microorganisms on the "Annotated List of Seed-borne Diseases" (Richardson, 1990) as justification for the proposed regulations (M. Condon, *pers. com.*).

There is an extensive world literature on seed-borne diseases that contains answers to many of these questions. However, it remains largely untapped, due to a lack of systematic organization of the information and to difficulties in accessing hard copy reference sources, particularly in developing countries. Recent advances in the electronic media present opportunities to disseminate information efficiently and economically throughout the world. An example is a new database on seed-borne diseases, produced by CAB International and the Seed Science Center, Iowa State University (Table 1) .

This database will provide:

- *An information source for effective management of seed-borne diseases.*
- *Enhanced capacity of users to identify seed-borne pathogens, assess their economic impact, and devise control strategies.*
- *Ready access to scientific information needed to make rational and justifiable decisions on plant quarantine regulations.*

Table 1. Structure of the CABI/ISU international database on seed-borne diseases.

Scope
- All major crops

Structure
- Potential economic impact of planting infected seeds
- Worldwide distribution
- Incidence of seed-borne infection
- Effect on seed quality
- Transmissibility of the pathogen by seeds
- Control by seed treatment
- Seed health assays
- Key references

Output
- Electronic and book format

Standardization of Seed Health Test Methods

The working sheets on seed-borne diseases published by the ISTA are good models for standardization of tests. However, only 68 sheets have been published over 40 years. Numerous new tests are needed to meet current and future demands for standardized tests.

Since 1994, the International Seed Health Initiative (ISHI) comprising industry and public scientists and regulators from Europe, Israel, Japan, and the US has been working on standardization of methods on economically important seed-borne pathogens of vegetables with a view to adding methods for field crop seeds at a later date. An agreement was reached in 1997 for ISHI to collaborate with ISTA to expedite the production of standardized seed health tests (Wesseling, 1997, Maddox, 1997).

Access to Worldwide Seed Treatment Regulations

Making current information available in on-line databases would greatly improve access to seed treatment and other phytosanitary regulations. An example of a regulatory database is the EXCERPT system, managed by USDA-APHIS and Purdue University (Anon., 1994). EXCERPT continually updates worldwide regulations that impact the export of US seed and other commodities.

Revision of Priorities in Seed Pathology Research

As discussed in the prefatory chapter of this book (McGee, 1997), seed pathology research needs to re-directed from an over-emphasis on cataloguing the presence of plant pathogen on seeds to the development of inoculum thresholds that will establish the economic risks of transmission of pathogens by seeds. Research in this area is fundamental to realistic and effective management of seed transmission of plant pathogens and little improvement in the worldwide seed health system can be expected unless priorities in seed pathology research are changed to meet this demand.

CONCLUSION

The overall goal of a world phytosanitary system is to protect against the spread of economically important pathogens without posing unnecessary barriers to worldwide movement of seeds. The present world phytosanitary system is not doing this effectively or efficiently. Resources are being misdirected to deal with unnecessary regulations while many potentially important seed-borne pathogens are being ignored. Implementation of the above four actions should lead to a system that is more effective in preventing spread of pathogens and less expensive to seed companies and governments throughout the world.

LITERATURE CITED

Anon. 1994. User's guide. Export Certification project (EXCERPT), version 1.2. USDA/APHIS and Center for Environmental Regulation Sytems, Purdue University. 176 pp.

Backman, P. A., Weaver, D. B., and Morgan-Jones, G. 1985. Soybean stem canker: an emerging disease problem. Plant Dis. 69:641-647.

Block, C. 1996. Biology of seed transmission of *Erwinia stewartii* in maize. Ph.D. Diss. Iowa State University, Ames, IA. 90 pp.

Kuan, T. L. 1988. Inoculum thresholds of seedborne pathogens: overview. Phytopathology 78:867-868.

Laszlo, E. M. 1982. Infection of soyabean varieties by bacterial blight (*Pseudomonas glycinea* Coerper) under field conditions. Novenyvedelem 18:443-447.

Maddox, D. 1996. Regulatory Needs for Standardized Seed Health Tests. Pages 81-92 in: *Plant Pathogens and the Worldwide Movement of Seeds,* ed. D. C. McGee. APS Press, St. Paul, MN. 109 pp.

McGee, D. C. 1988. *Maize Diseases: a Reference Source for Seed Technologists.* APS Press St. Paul, MN. 150 pp.

McGee, D. C. 1992. *Soybean Diseases: a Reference Source for Seed Technologists.* APS Press St. Paul, MN. 151 pp.

McGee, D. C. 1995. An epidemiological approach to disease management through seed technology. Annu. Rev. Phytopathol. 33:445-466.

McGee, D. C. 1996. Advances in seed treatment technology. APSA Technical Report No.11. 14 pp.

McGee, D. C. 1997. Relevance of seed pathology research priorities to worldwide movement of seed. Pages 1-16 in: *Plant Pathogens and the Worldwide Movement of Seeds,* ed. D. C. McGee. APS Press, St. Paul, MN. 109 pp.

Neergaard, P. 1977. *Seed Pathology,* Vols. I, II. John Wiley & Sons, NY. 1187 pp.

Richardson, M. J. 1990. *An Annotated List of Seed-borne Diseases, 4th Ed.* International Seed Testing Association, Zurich, Switzerland.

Sato, T., Viedma, L. Q, Alvarez, E., Romer, M. I., Morel P. W., and de-Viedma, L. Q. 1993. First occurrence of soybean southern stem canker in Paraguay. JARQ 27:20-26.

Wilcox, J, R. 1983. Breeding soybeans resistant to diseases. Pages 183-235 in: *Plant Breeding Reviews. Vol. 1.* ed. J. Janick AVI Publishing Co. Westport.

Wesseling, J. B. M. 1997. Government/Vegetable Seed Industry cooperation in the Netherlands. Pages 93-103 in: *Plant Pathogens and the Worldwide Movement of Seeds,* ed. D. C. McGee. APS Press, St. Paul, MN. 109 pp.

Yorinori, J. T. 1990. Stem canker of soybean. Comm. Tec. 44:1-8. EMBRAPA, Brazil.

CHAPTER 6

REGULATORY NEEDS FOR STANDARDIZED SEED HEALTH TESTS

D. A. Maddox

President
STA Laboratories
630 S. Sunset Avenue, Longmont, CO 80501

Seed health testing is an important component of a seed quality program. In the last 20 years, seed health has become an integral part of disease control for some economically important plant pathogens. For example, an outbreak of blackleg of cabbage, caused by the fungus *Leptosphaeria maculans*, in the eastern United States in 1973 was directly attributed to seed-borne inoculum from cabbage seeds produced in the northwest United States (Gabrielson *et al*, 1977, Maguire *et al*, 1978). Since the implementation of routine laboratory seed health tests for *L. maculans* in 1973, blackleg has been a rare occurrence in cabbage production areas (Fig. 1) (Gabrielson *et al*, 1977, Maguire *et al*, 1978).

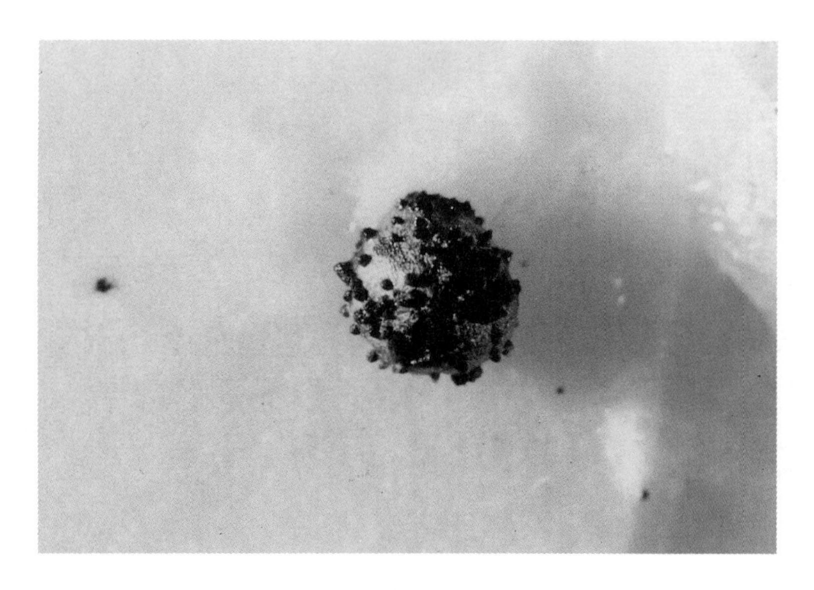

Figure 1. *Phoma lingam* pycnidia on a crucifer seed in a blotter test (Courtesy of Iowa State University).

Although seed health tests are important seed quality parameters and effective tools for managing disease risk, they cannot guarantee that a seed lot is free from a pathogen or that a crop will be free from a disease that may have been seed-borne. Seed health tests have the following goals:

- *Minimize losses to a seed-borne disease.*
- *Prevent introduction of a pathogen into an area by limiting seed-borne inoculum.*
- *Provide consistent and accurate detection of a pathogen in a seed sample.*

REGULATORY NEEDS FOR SEED HEALTH TESTS

As agriculture changes, there is an increasing demand for greater product quality and reliability. In the seed industry, this requires improvements and innovation in the processes for seed production. Today for example, transplants, high density plantings, and high vigor cultivars are more commonly used than in the past. It has also become clear that there has to be uniform and timely harvesting in order to meet "marketing windows" for companies to be profitable. These demands have, in part, been responsible for the development of reliable seed health tests as part of seed quality programs. Another demand has been the need for international seed shipments to meet phytosanitary regulations. Agriculture is a major issue in international trade agreements and in the era of the North American Free Trade Agreement (NAFTA) and the General Agreement on Trade and Tariffs (GATT), phytosanitary regulations are increasingly being used as a tools for countries to protect their agricultural commodities. As discussed by other authors in this monograph, phytosanitary requirements of some countries are not always based on scientific evidence or completely justifiable (Condon, 1997, McGee, 1997). Mexico, for example, had no phytosanitary requirements for import of vegetable seeds from the United States before 1991 but by 1994, phytosanitary regulations had been proposed for approximately 60 pathogens. A review of these regulations (unpublished report, American Seed Trade Association Seed-borne Disease Committee) revealed that ten pathogens were not seed-borne, eight already existed in Mexico, two are not known to be in the US, and eight have never been shown to be of economic importance anywhere in the world. This problem exemplifies the inadequacies and weaknesses in the current world phytosanitary system.

Phytosanitary regulations are negotiated between countries. The United States Department of Agriculture -

Animal and Plant Health Inspection Service (USDA-APHIS) is the agency responsible for phytosanitary issues in the US. The role of APHIS is also changing, however. The recent position paper, "The Seed Industry: A Background Paper" (USDA/APHIS, 1994), indicates that the agency will focus less on industry regulation and act more as a facilitator for US agricultural trade and economy. It will move toward cooperation with industry to establish self-regulation and take a stronger role in eliminating international pest regulations that act as trade barriers.

Since 1992, officials of APHIS and the American Seed Trade Association (ASTA) have been addressing concerns of the seed industry with respect to seed health testing, the role of private laboratories in phytosanitary testing, and phytosanitary requirements as trade barriers. Significant progress has been made toward establishing a US Seed Health Initiative (USSHI) with the following key elements identified:

- *International standardization of seed health test protocols.*
- *A system for laboratory accreditation in seed health testing.*
- *A scientific database for seed-borne diseases.*
- *An administrative structure for a national seed health program.*

STANDARDIZATION OF SEED HEALTH TESTS

Wide variation exists in methods currently used to test for economically important seed-borne pathogens (Tables 1 & 2). Four different methods can be used to detect *Xanthomonas campestris* pv. *campestris* (XCC), the cause of blackrot of crucifers (Table 1) including direct

Table 1. Examples of variation in seed health testing methods for particular pathogens.

Crop	Pathogen	Test Methods[a]	Citations
Bean	*Pseudomonas syringae* pv. *phaseolicola*	HI, IF, LP, PCR	Van Vurde *et al*, 1989
Brassica	*Xanthomonas campestris* pv. *campestris*	DP, IF, LP, PCR	Schaad, 1989; Randhawa *et al*, 1984
Cantaloupe	Squash mosaic virus	ELISA, GO	Nolan and Campbell, 1984
Celery	*Septoria apiicola*	BP, HI, SW/EX	Hewett, 1968
Corn	*Erwinia stewartii*	ELISA, GO	Lamka *et al*, 1991; McGee,, 1982
Soybean	*Phompsis* spp.	AP, BP	Kmetz *et al*, 1978; McGee *et al*, 1981
Tomato	*Clavibacter michiganense* pv. *michiganense*	IF/BIO, SW/LP, SG/LP	Fatmi and Schaad, 1988; Kritzman, 1991; Thry, 1969

[a]Abbreviations for test methods include: AP - agar plating; BIO - bioassay; BP - blotter; DP - direct seed plating; ELISA - enzyme-limked immunosorbent assay; GO - grow-out; HI - host inoculation; IF - immunofluorescence; LP - liquid plating; SW/EX - seed wash and spore examination; SG - seed grinding; PCR - polymerase chain reaction.

Table 2. Examples of the variation in seed sample sizes used in seed health testing methods for particular pathogens.

Crop	Pathogen	Seed Sample Sizes	Source
Bean	*Pseudomonas syringae* pv. *phaseolicola*	4,000-45,000	US seed laboratories
		5,000	Schaad, 1989
	Xanthomonas campestris pv. *phaseoli*	4,000-45,000	US seed laboratories
	Colletotrichum lindemuthianum	300-500	US seed laboratories
		200	Dutch seed laboratories
Brassica	*Xanthomonas campestris* pv. *campestris*	10,000-60	US seed laboratories
	Phoma lingam	10,000	Dutch seed laboratories
Cucubits	Squash mosaic virus	500-2,000	US seed laboratories
	Didymella brioniae	300-2,000	US seed laboratories
Lettuce	Lettuce mosaic virus	30,000	US seed laboratories
		2,000	Dutch seed laboratories
Tomato	*Clavibacter michiganense* pv. *michiganense*	10,000-40,000	US seed laboratories
		6,250	Dutch seed laboratories
	Xanthomonas campestris pv. *vesicatoria*	10,000-40,000	US seed laboratories
	Tobamoviruses	500-4,000	US seed laboratories
		2,000	Dutch seed laboratories

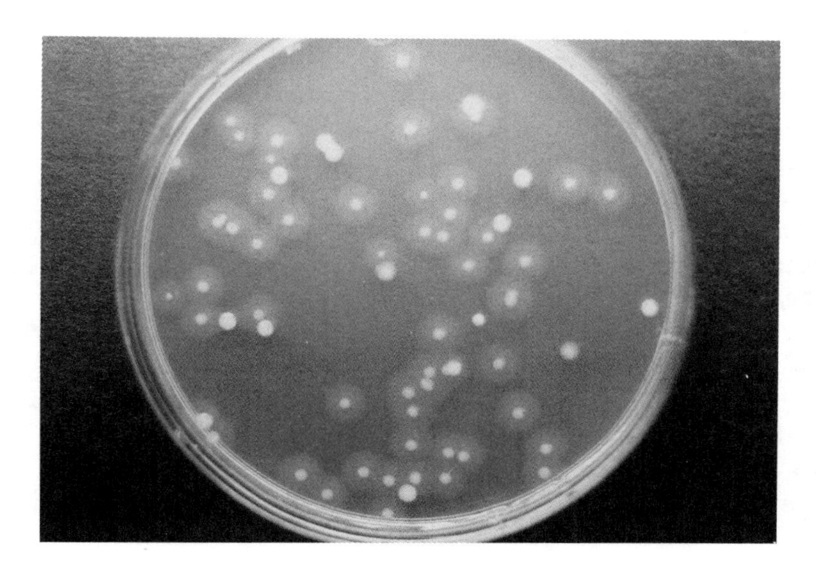

Figure 2. Selective medium for *Xanthomonas campestris* pv *campestris* showing typical halos (Courtesy of Seed Testing of America).

plating (Schaad, 1989), immunoflourescence (Schaad, 1978), liquid plating (Randhawa and Schaad, 1984, Schaad, 1989), and polymerase chain reaction (PCR) (R. L. Gabrielson, *pers. com.*). Furthermore, nine selective media are available for the detection of XCC (Fig. 2) and three extraction methods can be used in tests for *Clavibacter michiganense* subsp. *michiganense* (CMM) in tomato seed (Fatmi and Schaad, 1988, Kritzman, 1991, Thry, 1969). Sample size or quantities of seed to be tested can also vary greatly between tests. Table 2 lists the variation of seed quantities used in a range of seed health tests in laboratories worldwide. Because scientific studies relating sample size to inoculum thresholds (the level of inoculum of a pathogen in seeds that results in disease and economic loss) have rarely been published, the sample sizes listed (Table 2) result mainly from in-house studies and experiences of seed pathologists within private seed companies. Inoculum thresholds are influenced by the disease epidemiology in different agricultural areas and under different management practices. For example, a cabbage seed lot with one XCC infected seed in 10,000 seeds is more likely to cause economic damage in a transplant greenhouse in the southeast US, than if it is direct seeded in the dry climate of California under furrow irrigation. Other important considerations in the standardization of seed health tests include sensitivity of the assay, statistical reliability of the sample size, number of replications, sample uniformity, and prediction of disease risk.

Seeds need to be tested by internationally accepted methods that meet quality assurance and phytosanitary standards, thus obviating the need for multiple testing of the seed lot by industry and regulatory agencies. Although the means for standardization has yet to be established, criteria for a successful seed health standardization program would include:

- *Criteria for detection of seed-borne pathogens that includes test sensitivity, sample size, practicality, repeatability, and inoculum thresholds.*
- *Collection of available information on the protocols.*
- *Criteria for acceptance or denial of the method.*
- *Defined protocols and published methods.*
- *An administrative body to oversee test standardization.*

Achievement of an internationally accepted standardization system will require cooperation among seed groups and regulatory agencies (state, federal, and international) to determine the most scientifically sound methods for seed health testing.

ACCREDITATION OF SEED HEALTH TESTING LABORATORIES

A system for laboratory accreditation for seed health testing (USSHI) is the cornerstone of the cooperative efforts of APHIS and the US seed industry to improve the national seed health system. There are nine key elements to the laboratory accreditation plan:

- *Establishment and use of standardized seed health testing protocols.*
- *Establishment of a system for quality assurance in the laboratory.*
- *Document control with regular updates, identical copies, and limited distribution.*
- *Facilities and equipment to carry out testing.*
- *Documentation that equipment is working and calibrated.*
- *Training for personnel to ensure their expertise.*
- *Documentation of job descriptions and training requirements.*
- *Participation in a proficiency testing system.*
- *Audit and review system by an administration body.*

The USSHI accreditation of private and public laboratories for seed health tests would provide APHIS and the US seed industry with a credible system for phytosanitary testing comprising a group of testing laboratories that meet defined standards of reliability and quality. It would also provide seed companies with an alternative means of meeting phytosanitary requirements, which currently can only be obtained from a few public laboratories. Establishing accredited laboratories would provide other services to the regulatory agencies and the seed industry, such as the much needed source information on seed health protocols for economically important seed-borne diseases. Accredited laboratories would be an impartial source of data, standards, referee testing, and technical training for clinical application of seed health tests.

CONCLUSION

The higher demand for healthy seed as part of seed quality programs and the movement towards more restrictive phytosanitary requirements by many countries of the world has led the seed industry and regulatory agencies to re-evaluate the current world phytosanitary system related to seeds. There clearly are problems with the current system and without a systematic approach to their resolution, APHIS will be at a disadvantage in responding to unjustifiable phytosanitary requirements imposed by other countries. The international movement of US seeds may be impaired and significant increases in costs of production may result. The standardization of seed health testing protocols and the accreditation of laboratories for seed health testing would provide a means for reliable, consistent, uniform, and accurate information on the health of seed.

Finally, standardization and accreditation should provide a database of information on seed-borne diseases and seed health testing protocols. This database could be used to address phytosanitary requirements and provide sound scientific information for a justifiable, scientifically-based worldwide phytosanitary system. This information should enhance, not compromise, the current goals of the system to prevent the spread of economically important pathogens into new areas of the world.

LITERATURE CITED

Condon, M. 1997. Implications to international trading of seeds. Pages 17-30 in: *Plant Pathogens and the*

Worldwide Movement of Seeds, ed. D. C. McGee. APS Press, St. Paul, MN. 109 pp.

Fatmi, M. and Schaad, N.W. 1988. Semiselective agar medium for isolation of *Clavibacter michiganense* subsp. *michiganense* from tomato seed. Phytopathology 78: 121-126.

Frutchy, C. W. 1936. A study of Stewart's disease of sweet corn, caused by *Phytomonas stewartii.* Mich. Agri. Exp. Stn. Tech. Bull. 152.

Gabrielson, R. L., Mulanax, M. W., Matsuoka, K., Williams, P. H., Whiteaker, G. P., and Maguire, J. D. 1977. Fungicidal eradication of seed-borne *Phoma lingam* of crucifers. Plant Dis. Rep. 61: 118-121.

Hewitt, P. D. 1968. Viable *Septoria* spp. in celery seed samples. Ann. Appl. Biol. 61: 89-98.

Kritzman, G. 1991. A method for detection of seed-borne bacterial diseases in tomato seed. Phytoparasitica 19: 133-141.

Lamka, G. L., Hill, J. H., McGee, D .C., and Braun, E. J. 1991. Development of an immunosorbent assay for seed-borne *Erwinia stewartii* in corn seeds. Phytopathology 81: 839-846.

Maguire, J. D., Gabrielson, R. L. Mulanax, M. W., and Russel, T. S. 1978. Factors affecting the sensitivity of 2,4-D assays of crucifer seed for *Phoma lingam.* Seed Sci. Technol. 6: 915-924.

McGee, D. C. 1982. Seed disease testing in Iowa. Iowa Seed Sci. 4: 3-4.

McGee, D. C. 1997. World phytosanitary system: problems and solutions. Pages 67-79 in: *Plant Pathogens and the Worldwide Movement of Seeds,* ed. D. C. McGee. APS Press, St. Paul, MN. 109 pp.

McGee, D. C., and Nyvall, R. F. 1981. Soybean seed health. Coop. Ext. Serv. Iowa State Univ. Pm 990.

Nolan, P., and Campbell, R. N. 1984. Squash mosaic virus in individual seeds and seed lots of cucurbits by enzyme-linked immunosorbent assay. Plant Dis.68: 971-975.

Randhawa, P. and Schaad, N. W. 1984. Selective isolation of *Xanthomonas campestris* pv. *campestris* from crucifer seeds. Phytopathology 74: 268-272.

Schaad, N. W. 1978. Use of direct and indirect immunoflourescence tests for identification of *Xanthomonas campestris*. Phytopathology 68: 249-252.

Schaad, N. W. 1989. Detection of *Xanthomonas campestris* pv. *campestris* in crucifers. Pages 68-75 in: *Detection of Bacteria in Seed and Other Planting Material*.eds. A.W. Saettler, N. W. Schaad, and D. A. Roth, APS Press, St. Paul, MN. 122 pp.

Thry, B. D. 1969. Assaying tomato seed for *Corynebacterium michiganense*. Plant Dis. Rep. 53: 858-860.

USDA/APHIS 1994. *"The Seed Industry: A Background paper"* Policy and Analysis Department, United States Department of Agriculture, Animal and Plant Health Inspection Service, October, 1994.

Van Vuurde, J. W. L. and van den Bovenkamp, G. W. 1989. Detection of *Pseudomonas syringae* pv. *phaseolicola* in bean. Pages 30-40, in: *Detection of Bacteria in Seed and Other Planting Material*. eds. A.W. Saettler, N. W. Schaad, and D. A. Roth, APS Press, St. Paul, MN. 122 pp.

CHAPTER 7

GOVERNMENT AND VEGETABLE SEED INDUSTRY COOPERATION IN THE NETHERLANDS

J. B. M. Wesseling

Head Inspections and Analysis Section
NAKG
Sotaweg 22, P.O. Box 27, 2370 AA Roelofarendsveen,
The Netherlands

More than 50 years of cooperation between the inspection system of NAKG (General Netherlands Inspection Service for Vegetable and Flower Seeds) and the vegetable seed industry in the Netherlands has produced an infrastructure that guarantees high quality seed and meets all legal requirements of the Netherlands. NAKG uses the quality control systems of the seed companies as much as possible and, in turn, strengthens and supports the quality management systems of the companies.

This article will describe the activities of NAKG and the basis for the cooperation between NAKG and the vegetable seed industry in the company accreditation system. It will also outline NAKG's national seed-borne pathogen program in the Netherlands and its collaboration with international organizations.

ACTIVITIES OF NAKG

The main activities of NAKG are summarized on Figure 1. The official tasks are done on behalf of the government of the Netherlands and on the basis of accreditation partly or fully delegated to the seed companies. Services are also provided to companies by NAKG. A short explanation of these activities is given.

Official Tasks

These are described as follows:

Inspection - Inspect vegetable seeds and plants, flower seeds and seedlings, onion sets and seed shallots for germination, varietal trueness and uniformity, and plant health.

EU Plant Passports - Provide phytosanitary inspection and certification for quarantine diseases of propagation material. Almost 100,000 plant passports are issued per year, which allow for movement of material between EU countries. Propagative material without quarantine diseases does not require a plant passport.

European Description Services - List and describe vegetable and flower varieties.

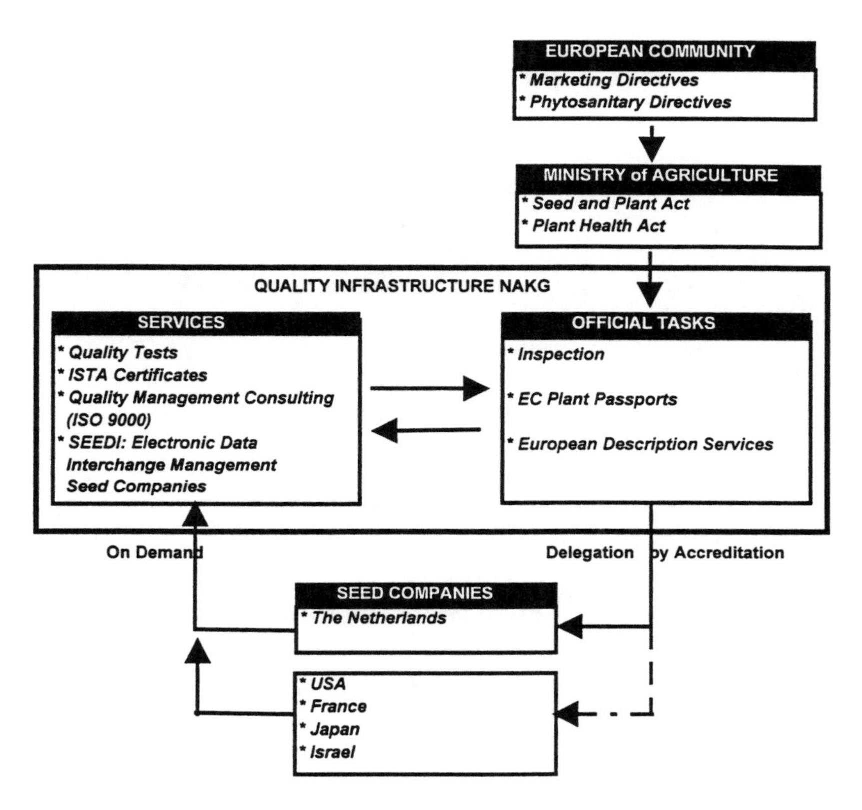

Figure 1. Activities of NAKG as they relate to governmental bodies and to seed companies.

Services

The following services are carried out by NAKG.

Quality Test - Provide tests for germination, seed health, and physical purity.

ISTA Certificates - Issue certificates for germination, seed health, and physical purity, based on tests that comply with rules of the International Seed Testing Association.

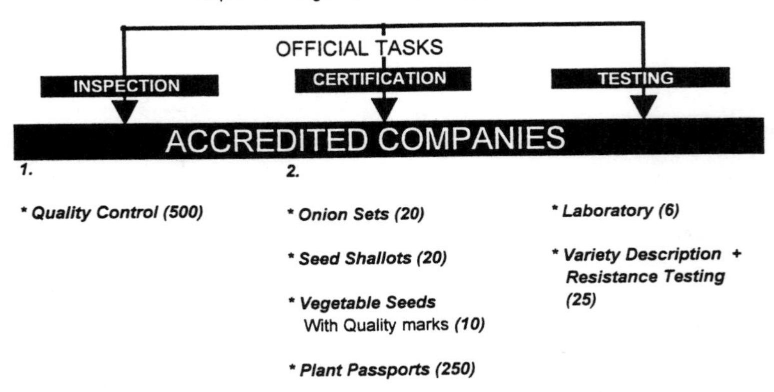

1. Costs for Inspection Annually
Inspection based on Accreditation: 0.2 - 0.5% of the yearly turnover.
Inspection based on Product Certification: 5 - 10% of the yearly turnover.

2. Annual savings in certification tasks resulting from accreditation
Dfl 1,000,000 (US $500,000)

Figure 2. Inspection, registrations, and seed testing programs of NAKG.

Consultancy - Provide consultancy on total quality management principles and systems based on ISO 9000 to seed companies, flower bulb companies, plant raisers, and growers.

Training - Train personnel in seed companies on many different subjects.

SEEDI - Provide consultancy on management/coordination of the Electronic Data Interchange System for the vegetable seed industry. This is a new consulting activity added to the services of NAKG at the request of the seed companies.

ACCREDITATION OF SEED COMPANIES

The cooperation between NAKG and a seed company is founded on an accreditation system (Figure 2). A company can be accredited after it is determined that it can fulfill certain requirements. These requirements relate to the quality care system and control of the production process of the company. Elements such as quality policy, control of critical points in the process, organization of the quality control, tasks and responsibilities of staff, and training of staff are part of the requirements. Accredited companies have to be checked annually by an independent body. Accreditation guarantees that the company controls its processes, but it does not determine the level of the quality of the product. The quality of the product depends on the objectives of the company or the arrangements between the customer and the company.

An accreditation system is an efficient and effective way of controlling quality for the following reasons:

- *The companies have a major economic stake in ensuring the quality of their products. Therefore, they should be trusted to take the responsibility for quality control.*
- *The companies can most effectively perform quality control because they have the best knowledge of their own material.*
- *The system is efficient because retesting by the inspection services is no longer required.*
- *New quality control elements can be incorporated very quickly, efficiently, and effectively.*

In the past, official sampling, testing, labeling, and inspection were carried out by the NAKG or the Netherlands Plant Protection Service and paid for by the seed companies. The accreditation system allows this work to be done by the companies. This has resulted in savings of approximately US$500,000 per year. Another beneficial effect is the logistical freedom of the companies, in that they do not have to wait for the inspection service to test the seeds before they receive certificates.

In short, the cooperation between the NAKG and the companies ensures an optimal result with respect to marketing performance, quality care for the clients, and public control obligations of the Dutch seed law and EU-directives.

SEED-BORNE PATHOGEN PROGRAM

Seed-borne pathogens have been a particularly active area of cooperation between seed companies and the NAKG. This was motivated by a problem resulting from contamination of pepper seed by pepper mild mottle virus. The problem was confounded by the lack of a standard seed health test protocol and different test results obtained by the official inspection bodies and research institutes. The investigation was complex, lengthy, and expensive to the companies involved.

In the Netherlands it was perceived that the problem might have been prevented if there had been an internationally-accepted quality assurance system for the control of this seed-borne pathogen. Realizing that similar problems could develop for other seed-borne pathogens, Dutch companies organized within NVZP (Dutch Seed Trade Association) and NAKG decided to initiate the Seed-borne Pathogen Program.

This program has the following objectives:

- *Develop standardized protocols to test for seed-borne pathogens.*
- *Develop a national system for accreditation of companies which assures that tests done with standardized protocols would be accepted by other companies, customers, and officials.*

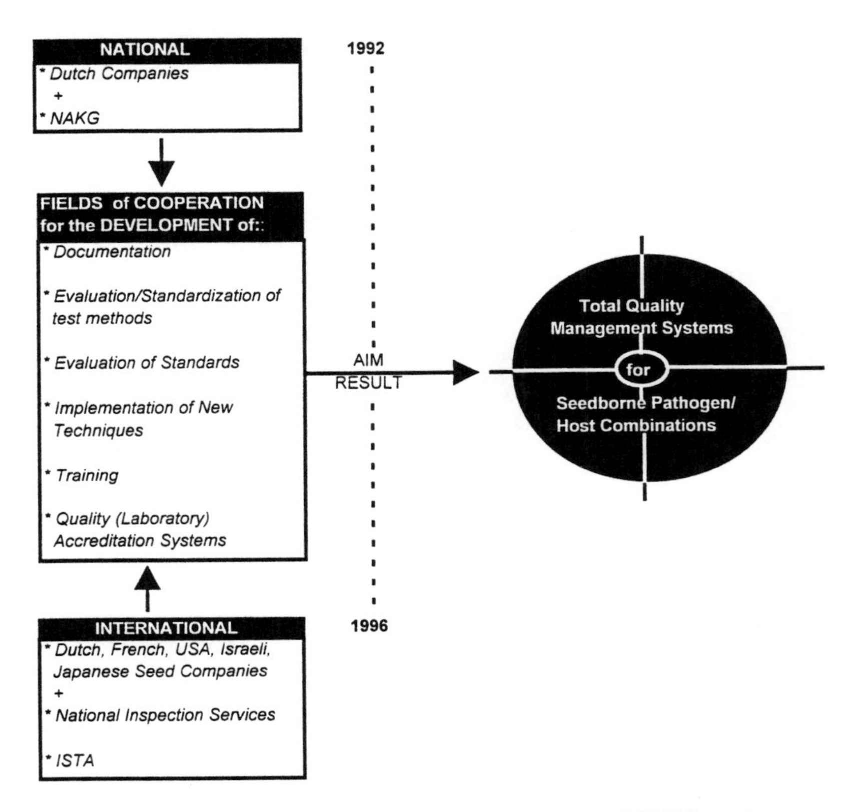

Figure 3. The seed-borne pathogen program at NAKG as it relates to national and international seed health organizations.

- *Minimize retesting by the government, seed company or independent testing laboratories that has been necessary due to concerns about the validity of the risks connected with unstandardized testing.*

Progress on the program since its commencement in the autumn of 1992 is summarized in Figure 3.

Development of Documentation

A risk analysis system was used to prioritize 35 economically important host/pathogen combinations from a total of 300. A standardized questionnaire was then circulated among the participants of the project in order to assemble a body of information on each of the 35 pathogens. The questionnaire solicited specific information on pathogen characteristics, disease risks, economic impact, seed health test methods and standards, disease control measures, and seed treatments. Detailed documentation has now been obtained for more than 20 host/pathogen combinations.

Evaluation of Test Methods

For each selected host/pathogen combination, the status of seed health test methods was analyzed as follows:

Test method available but not reliable- Seed-borne diseases for which the specificity toward the pathogen is not accurate enough. To more accurately identify the pathogen, further taxonomic analysis is necessary. An example of this category is *Fusarium solani* in squash and pumpkin seeds.

Test method not available- Seed-borne pathogens for which no routine-testing method is available, research programs will be initiated to develop a new test method. An example of this category is cucumber green mottle virus in cucumber seeds.

Evaluation of standards- Interpretation of test results to determine whether a seed lot is accepted or rejected has often been arbitrary for many host/pathogen combinations because of the lack of well defined standards. Sample size and sensitivity of the method are important criteria in determining accurate standards. Records of seed testing information from the seed companies and NAKG will be used to generate epidemiological information on the selected pathogen/host combinations. This approach will facilitate a means to determine the incidence of seed infection that will cause disease problems in the field. If necessary, a research project will be developed to obtain information needed to set the standards.

Implementation of new testing methodology - The number of very specific and sensitive detection techniques based on biotechnology is rapidly increasing. Several seed companies have invested in research on techniques such as PCR, RAPD, and RFLP to detect seed-borne pathogens and to discriminate between saprophytic and pathogenic isolates. The seed-borne pathogen program will invest in research to determine the feasibility of this new seed testing methodology. The new methodologies will only be implemented if they improve the existing methods.

Training programs - Training of personnel is a very important component of an effective quality control system. NAKG and seed companies will develop and organize:

- *Training for staff of individual seed companies.*
- *Courses for new test methods.*
- *Workshops to update knowledge of seed-borne diseases and seed health tests.*

Implementation of an accreditation system for seed health laboratories - The accreditation process is separated into two parts. The first component comprises general accreditation of the laboratory. Requirements for this are based on ISO-Guide 25 and concerns laboratory equipment, staff knowledge, the quality system, administration, and operating system etc. The monitoring of this system is done annually through a complete assessment.

The second part regulates the tests for which the accreditation has been granted and the test protocols which are required. If the method used by a laboratory is different from the standard method, it is compared with the standard method and accepted if proven to be reliable.

Accredited laboratories can only use standard methods or their own methods, if approved. This part of the system is monitored by proficiency tests and workshops. A small number of post-control samples from marketed seed lots are tested by the NAKG for comparison with the results of the accredited laboratories in order to secure the system.

At present, the accreditation system is concentrating on laboratory test methods. Seed-borne pathogens can also be controlled by field inspections, seed treatments or other measures. At a later stage, it is intended to extend the accreditation to include the complete quality control system for a host/pathogen combination.

Internationalization of the System

Initially the accreditation system functioned at a national level, but the final goal is for an international system. In a meeting in the Netherlands in February 1994, it was decided to upgrade this project to an international level involving French and US vegetable seed companies (Fig. 3). A second meeting in Angier, France in December 1994, began the working phase of the international effort. A third meeting in Ames, Iowa, US in October 1995 resulted in the inclusion of Israeli and Japanese companies. In 1995, the project was named the International Seed Health Initiative (ISHI) at a joint meeting with the International Seed Testing Association - Plant Disease Committee (ISTA-PDC) meeting in Cambridge England. This meeting also led to an agreement to develop a close collaboration between ISHI and ISTA on seed-borne disease issues and to integrate activities where possible. This is a unique effort in which seed companies from different countries are cooperating with the national inspection bodies to realize a harmonized worldwide seed health testing system.

CONCLUSION

Accreditation of seed companies for the quality control of seeds guarantees that legal aspects of the trade of seeds are covered in a reliable manner. This system is very efficient in reducing inspection costs for companies but also benefits the production process of the companies as a whole. Advantages to the inspection service are that new developments in quality control of seed and new legal obligations can be incorporated into the system very quickly. After more than 50 years working with seed companies, the experience of the NAKG has produced an accreditation system founded on the following principles that has benefits to both parties:

- *Do it once, do it well*
- *Confidence in competence of companies*

LITERATURE CITED

ISO - Guide 25. 1991. General requirements for the competence of calbration and testing laboratories. (3rd ed., iggo). International Standards Organization, Geneva, Switzerland.

ISO - Guide 9004-1. 1994. Quality management and quality assurance standards. Part 4: Guidelines for selection and use (1994-07-01). International Standards Organization, Geneva, Switzerland.

ISO - Guide 9001. 1994. Quality systems - model for quality assurance in design development, production installation, and servicing. (2nd ed., iggu -07-01). International Standards Organization, Geneva, Switzerland.

INDEX